Environmental Education
for the 21st Century

PETER LANG
New York • Washington, D.C./Baltimore • Boston
Bern • Frankfurt am Main • Berlin • Vienna • Paris

Environmental Education for the 21st Century

International and Interdisciplinary Perspectives

Patricia J. Thompson, Editor

in cooperation with the Office of Media Relations
and Publications at Lehman College
Anne D. Perryman, *Director*

with an Introduction
by Ricardo Fernández,
President of Lehman College, City University of New York

PETER LANG
New York • Washington, D.C./Baltimore • Boston
Bern • Frankfurt am Main • Berlin • Vienna • Paris

Library of Congress Cataloging-in-Publication Data

Environmental education for the 21st century: international and
interdisciplinary perspectives/ edited by Patricia J. Thompson.
p. cm.
Includes bibliographical references and index.
1. Environmental education—Congresses. 2. Environmental policy—
Congresses. 3. Environmental ethics—Congresses. 4. Ecofeminism—
Congresses. I. Thompson, Patricia.
GE70.E5817   333.7'071—dc21   97-24484
ISBN 0-8204-3749-2

Die Deutsche Bibliothek-CIP-Einheitsaufnahme

Thompson, Patricia J.:
Environmental education for the 21st century: international and
interdisciplinary perspectives/ ed. by Patricia J. Thompson.
–New York; Washington, D.C./Baltimore; Boston; Bern; Frankfurt am Main;
Berlin; Vienna; Paris: Lang.
ISBN 0-8204-3749-2

Cover photograph by Barbara P. Cardillo.
Cover design by Goodspeed Associates.

The paper in this book meets the guidelines for permanence and durability
of the Committee on Production Guidelines for Book Longevity
of the Council of Library Resources.

© 1997 Peter Lang Publishing, Inc., New York

Printed in the United States of America.

Dedication

This volume is dedicated to

Theresa Lato

and

Margaret Ruestow,

whose commitment to environmental activism

and education has inspired generations of educators.

*E*verybody talks about the environment.

*N*aturally.

*V*enomous machines and toxic schemes of

*I*ndustry gone mad spoiling everything,

*R*ushing into news like somebody accidentally hit fast-forward.

*O*ur flora, fauna, feathered flight in danger of demise…

*N*o one wants to sacrifice conveniences, elusive free time.

*M*aybe someone else will keep an eye out, while we sleep. Oh yes,

*E*verybody jaws about this precious environment

*N*ow and then. Naturally.

*T*alk is cheap.

—Kate Potter
Summit Station, Pennsylvania

Contents

Part I: A Global Perspective

Chapter

Introduction

From Ivory Tower to Green Tower:
A President's Perspective

Ricardo R. Fernández
Lehman College
The City University of New York

The view from my office is of a small, calm body of water surrounded by trees. It reminds me of Wisconsin, with its many lakes and rolling, deep-green fields, where I lived for more than 25 years. But I am in the northwest corner of the Bronx, the northern borough of New York City. And while the view from my vantage is bucolic, there are many reminders of the densely-populated neighborhood that surrounds us: high-rise buildings nearby, noisy traffic below, and the occasional jetliner overhead on its way to or from LaGuardia Airport.

The 37-acre tree-lined campus of Lehman College is isolated on three sides; only to the south is there a contiguous building—Walton High School, separated from the college by a wrought iron fence. To the north is Harris Park, with its city-owned athletic fields, lined with majestic sycamores. To the east and deep below the street, hidden by thick bushes, lie the New York Transit Authority's Bedford Yards. There subway cars are repaired and serviced. Along the college's western boundary lies the 195-acre Jerome Park Reservoir which, to this day, supplies about ten percent of New York City's fresh water. It is this view that I am privileged to enjoy. It is a soothing environment. Our students speak of it as peaceful, serene—an "oasis."

Surprisingly, the Bronx is ideally suited for environmental study. As three early historians of the borough have noted:

> [T]he raw data for our thinking lies under our feet and constitutes the environment on all sides of us. The path along which we walk, the ground through which the path is cut, the flow of the river, the rustle of the trees, the configuration of the land, the distant peak, the nearer

palisades—all these have meanings…. The earth speaks to us with a voice as familiar and comprehensible as the conversation of the friend with whom we drink our coffee. (Wells, Haffen & Briggs, 1927, 1:1) The situation, I believe, is much the same seventy years later.

Reflections on Bronx History

The earliest history of the Bronx is also the earliest history of New York City. Giovanni Verrazano in the 16th century and Henry Hudson in the 17th were among the first to explore the area. New York, of course, was first known as New Amsterdam. One of the early settlers in the northern vicinity in these early years was a Dutch farmer, Jonas Bronck, who purchased more than 500 acres of land between the Harlem and Acquahug (later "Bronx") rivers. The story is told of early New Yorkers who enjoyed weekend outings at "the Broncks' " farm (Wells, Haffen & Briggs 1927, I:19). Today the Bronx, a natural derivative from the early name, is a borough, as New York City administrative units are called.

The Bronx has always been multicultural. Settled before the American Revolution by Dutch, French, and English settlers, it has retained its identification with the earliest Indian settlers: Algonquin, Mohawk, and Iroquois. It still bears local names from its early history: Spuyten Duyvil (Dutch), Kingsbridge (English crown), and Mosholu (Indian).

What is now the Bronx grew merely as the most southerly part of a vast stretch of county territory. It received the overflow of Manhattan. It was a land of adventure for the first daring and curious spirits that crossed the Harlem stream to make a home deep in the woodland and deep in the wilderness. (Wells, Haffen & Briggs, I:309)

Later a "bedroom community" for New York City workers, the Bronx has been a distinctive administrative unit from its early settlement. Annexed to New York City in 1874, the population grew rapidly and necessitated building the Jerome Park Reservoir, begun in 1895 on land that was once a racetrack. Today the Bronx is home to 1.2 million people.

Urban Ecology and the Campus Setting

Lehman College of The City University of New York (CUNY) is a four-year liberal arts institution located in a cluster of leafy neighborhoods in a borough that has served—unfairly, I should point out—as a national metaphor for urban blight and decline. In fact, the Bronx has more parkland than any other borough of New York City and many of the largest open spaces remaining in the city. Within a few minutes of campus are the New York Botanical Garden, the New York Zoological Society (Bronx Zoo), Van Cortlandt Park (second only to Central Park in size), the Wave Hill environmental center, and a number of exclusive residential neighborhoods along the Hudson River. Indeed, even the hard-pressed community of the South Bronx—where the negative image was created—has been largely rebuilt in the last two decades. The borough is also the site of important medical and health care facilities, notably Montefiore, Bronx Lebanon hospitals, and the Albert Einstein College of Medicine. While Lehman College—like all of the CUNY colleges—serves many students who live in economically disadvantaged neighborhoods, it is—thanks to these students and to our distinguished faculty—a socially vibrant and intellectually vital community.

An Updated Version of Town and Gown

Colleges and universities have always had a relationship to the surrounding areas. Today, "surrounding area" has global implications. As Warren Bennis (1996) warned, "Around the globe, humanity currently faces three extraordinary threats; the threat of annihilation as a result of nuclear accident or war, the threat of a worldwide plague or ecological catastrophe, and a deepening leadership crisis" (154). Thirty years ago, ecologist Garrett Hardin pointed to the "tragedy of the commons" as a way to sound the alarm about the impending dangers of taking our environment for granted. The "commons" today is an even more critical issue because we live with a growing consciousness of global environmental concerns. With an eye on the coming millennium, it seemed prudent to bring some of the leading environmental scholars and activists into conversation on our campus.

Our efforts to heighten environmental awareness led to the designation of the 1994-95 academic year as "Earth Year at Lehman College." The culmination of this year-long theme was the international conference "Environmental Issues for the 21st Century," held in April 1995 to commemorate the 25th anniversary of Earth Day. The conference was designed to encourage multidisciplinary approaches to environmental topics, to increase the broader community's awareness of environmental issues, to stimulate collaboration between public and private sectors in formulating environmental policy, and to recognize individuals and organizations for their con-tributions to improving the environment of the Bronx. The college com-munity was challenged to bring its various divisions, departments, pro-grams, and services—including the Library, Art Gallery, and computer fa-cilities—into this week-long, campus-wide academic event. The poem "En-vironment," by Kate Potter, which opens this volume, was written espe-cially for this event (see Perryman, this volume).

A President's Predicament

Can an institution of higher education, particularly one in a large city, play a leadership role in the promotion of awareness and action-oriented initia-tives to improve both the campus environment and that of the surrounding community? As the president of the only public senior college in the Bronx, I asked myself this question soon after arriving at Lehman College in the Fall of 1990. I felt that a necessary first step was to begin at home, so to speak, by ensuring that our own busy campus be well-maintained and free of litter—and to emphasize that this was a job for all of us, not merely the work of the Buildings and Grounds staff. To the astonishment of adminis-trators, faculty, staff, and students, when I saw a candy wrapper, sandwich bag, or soda can left behind by others, I picked it up personally and depos-ited it in a trash container. My purpose was to change the culture of the institution—to incorporate into its value system a sense of responsibility of each member of the college community (including the president) to main-tain the campus litter-free.

Many areas in the Bronx, including neighborhoods near our campus, were strewn with debris and an occasional tire or two (left behind by some-one who wanted to avoid paying the disposal fee). Fortunately for Lehman College (and coincidental to our efforts), Fernando Ferrer, the Bronx Bor-ough President, sued New York City in 1994 to improve garbage collection and street sweeping by the Department of Sanitation. In a matter of weeks the response was dramatic; major thoroughfares were cleared of accumu-lated debris, and garbage collection became more consistent throughout the borough. We could see the direct result of a leadership that refused to allow a lower standard for residents of the Bronx than for people who live in Manhattan.

Lehman College employees and students—through the President's "Green Team"—had been collecting recyclable white paper, cans, and bottles long before this was mandated by law. Sharing responsibility for campus environmental goals brought the college community together—for this issue at least—and demonstrated to a beleaguered Buildings and Grounds staff that their efforts were neither ignored nor in vain. To be effective, of course, such support must be ongoing. The emphasis on individual responsibility has given our students a sense of pride and a feeling of ownership in the attractive appearance of their school's buildings and campus. When students are asked why they chose Lehman from among the senior colleges in the City University system, a surprisingly large number say, "the beautiful campus."

The Impact on the Community

All these activities, if limited to the campus, however, do not have the impact they might have, nor do they demonstrate to our neighbors the importance of the college to local concerns and the support that can be forthcoming from an institution of higher learning when it is integrated into its surrounding community in a pragmatic way. Two specific projects at Lehman illustrate the college's involvement in environmental issues beyond the perimeter of the campus.

Adopt-a-Station

Lehman College was the first organization in the Bronx to join the Transit Authority's "Adopt-a-Station" program. Campus volunteers "adopted" the subway station a block from the eastern entrance to the college, through which thousands of students and visitors pass yearly on their way to classes and special events. This old run-down station, adjacent to a bridge that was closed to vehicular traffic in 1991, was transformed through this initiative. In cooperation with the station manager, the building was freshly painted; volunteers built benches and planted flowers. Merchants took note of the change and began to ask: What's happening? Who is responsible? How can we help? Nearby residents began to see the improvements; some of them called or wrote us to share their positive feelings. As a result of the college's effort, the Transit Authority put the station into a 24-hour graffiti-free program.

The Jerome Park Conservancy

Another example of the College's involvement in the community environment is our participation in the Jerome Park Conservancy, a newly incorporated entity dedicated to the improvement of the public green space surrounding the Jerome Park Reservoir. This New York City reservoir, which is part of the Croton reservoir system to the north, is currently under a Federal mandate to filter its water by the early years of the next century. In 1994, the New York City Department of Environmental Protection unveiled a plan to build a huge water filtration plant in the northern section of the Jerome Park Reservoir. Given the project's potentially detrimental impact on the surrounding community—including the Lehman campus—many local residents began to organize in opposition to the filtration plant.

The direct impact of such construction on the Lehman College campus would be less than on some of the nearby residential areas, as well as two high schools across the street from the site of the proposed filtration plant. Nevertheless, I joined the Conservancy in its efforts to oppose the plant's location in this densely populated area and, early in 1995, became Chairman of its Board of Directors. Beyond opposition to the construction of the filtration plant at this site, the Conservancy was also committed to developing and promoting an alternative vision of what might be possible for the reservoir. The work of the Conservancy led to the development of a comprehensive plan to build a public park around the Jerome Park Reservoir instead of the filtration plant, and many residents from the surrounding community embraced it. Although the final decision on the location for the proposed filtration plant will not be decided for several years, the collaborative relationships that have evolved between members of the Conservancy and Lehman College faculty and staff should lead to even greater interaction between the college and the community in the years ahead. We have put an "environmental spin" on traditional notions of "town and gown."

Building Environmental Partnerships

In both cases—with the Adopt-a-Station program and the Jerome Park Conservancy—Lehman College has served as a partner of a project of vital interest to community residents. In the first instance, Lehman staff and students initiated the idea, pursued it, and brought it to fruition with the help of elected officials and the Transit Authority. In the second, community activists invited the president of the college to join their effort, and this led to the involvement of other members of the college staff. Students and faculty have shown their concern over the proposed filtration plant and have received regular briefings at College Senate meetings, general faculty meetings, and in other forums. As the time approaches for a decision to be made, even greater involvement by Lehman students and the nearby public high schools is likely to occur.

The Impact on Curriculum

The most direct and lasting impact a college can have on its students is delivered through the curriculum. One outcome of our environmental initiative was the proposal for a new program in Environmental Studies and Policy, developed by faculty members in the 1994-95 academic year. This program, which involves the Departments of Biological Sciences, Chemistry, Physics, Geology and Geography, Psychology, Political Science, Sociology, as well as English, Languages and Literatures, Philosophy, and Education, has an emphasis on policy issues. Its aim is to provide students with a broad, liberal arts background that will allow them to study and analyze environmental topics from various disciplinary perspectives (see Wille, this volume). Following graduation, these students will be able to work in a variety of fields, such as elementary education, journalism, policy analysis, and also pursue graduate study in an area of their choice, such as education, law, medical and health studies, political science, psychology, or sociology.

Looking Ahead

Although I will refrain from predicting the future as the end of the millennium approaches (many others will do the honors), it is not foolhardy to imagine that, in the next century, environmental issues will take on critical importance in the public eye. It is appropriate that colleges and universities, which have been the centers of activism regarding the environment since the first Earth Day celebration in 1970, continue to be centers of debate and consciousness-raising. We have just begun to study the environment through the disciplines of philosophy, economics, and political science with a focus on public policies and practices. Thus, books such as this can serve an important function by providing the perspectives of scholars from several countries on the education of the public and of school-age youth about the environment. As we educate the young about their environment, we must instill in them an ethic of caring and responsibility for the planet—so that every person does his or her share to use resources wisely and to preserve our natural riches for future generations. This may be a slow process, but ultimately it will be the best and only way to bring about permanent change. As one environmental educator has observed:

> The web of life is a key environmental concept. We also learn to value the habitats of each creature and the critical role each member of a community plays in maintaining a viable ecosystem. As the creatures of non-human systems all depend on each other, so do we creatures of the human systems depend on each other. It is only when we develop and inculcate the same respect for human habitats as we do for those of fish and birds and mammals that we will begin to effectively address living in harmony with the non-human organisms of this planet. (Cohen 1994, 1)

We have tried to bring this perspective into the culture of our college.

Lehman College, CUNY
The Bronx, New York
February, 1997

References

Bennis, Warren. 1996. "The Leader as Storyteller." *Harvard Business Review* (January/February), 154-60.

Cohen, David L. 1994. "Valuing Human Habitats." *EEAC Newsletter* (Spring), 1.

Hardin, Garrett. 1968. "The Tragedy of the Commons." *Science 1*:162 (December 13), 1243-48. Adaptation of an address given June 25, 1968.

Wells, James L., Louis F. Haffen & Josiah A. Briggs. 1927. 4 vols. *The Bronx and Its People: A History 1609-1927.* New York: The Lewis Historical Publishing Co.

Notes of a Green Pencil Editor

Patricia J. Thompson
Lehman College
The City University of New York

This volume's genesis was the announcement in 1993 that Lehman College would mark not just the 25th anniversary of Earth Day in 1995 but that the entire 1994-1995 academic year would be marked by environmental themes and activities. The year's culminating event (during Earth Week in April) was a week-long conference with a variety of events aimed at raising the environmental consciousness of students, faculty, and administration. The goal was to recognize the systemic interdependence of the campus and its surrounding community. As chair of the conference planning committee and the conference convener, I come to this volume with an appreciation of the broad range of interests that environmental study encompasses. My particular perspective comes from human ecology with a focus on the impact of the environment on the wellbeing of individuals, families, and households, as well as their impact on the environment. This includes the way households and communities manage their resources—natural, human, and humanmade—to maintain an optimal quality of life. In my own language, this is a Hestian perspective, an adjective derived from Hestia, the ancient Greek goddess who symbolized the daily round of human activity.

Background

When we look at the spectrum of current academic offerings in most college bulletins, green (as a metaphor for environmental study) is not uniformly diffused throughout the curriculum. For some time, those with a commitment to green studies have had to climb a mountain of resistance, swim rivers of indifference, drift in oceans of misinformation, and cut through forests of confusion before they find themselves in a place that is hospitable to viewing school subjects and academic disciplines through a green lens—a "green lens" that can be applied not just to academic study but to the everyday lives of a knowledgeable citizenry. At least, that is how I felt as I struggled to bring an environmental perspective into my teaching and writing. Based on two decades of experience and observation, I am convinced that people can make a difference and that environmental decision-making must begin with raising environmental awareness at home and in the community. This was brought home to me as early as 1973 when, as a result of the Arab oil embargo, a group of Lehman College students, under my supervision, conducted a National Science Foundation supported, student-originated study, "The Energy Crisis and Family Decision-Making."

If we are to be effective at policy-making levels, we cannot forget the complex issues surrounding human motivation and behavior. While recognizing the fundamental role of the natural sciences in providing the knowledge base on which sound policy must be predicated, we can also draw upon the human and social sciences to assess other factors in environmental studies. I recall that, some 20 years ago, the editor of one of my secondary school textbooks referred to my environmental perspective as a "fad." I told her then, and I would tell her now, that if ecology is a fad, it is the *last* fad! Rapid and complex changes in society and technology create new human/environment interfaces that must be studied by the social, family, and consumer sciences. With an interdisciplinary approach, we can address such interrelated issues as air quality, water purity, soil erosion, noise pollution, pesticides in the food chain, deforestation, toxic waste, and the role of environmental conflicts and the arms race.

Overview

This volume opens with Joy Palmer's call for a global perspective in environmental study. She underscores the planetary imperatives for education, public policy, and activism. She argues that EE is not the sole province of science educators. Gaining an ecological perspective involves the cooperation and collaboration of all disciplines and channeling accurate information into classrooms and laboratories at various levels in the educational system as well as to the public. Jacqueline Gravanis takes us back to Greek myth as the starting point of our evolving concepts of humankind's place in nature. She addresses the current strategies employed by environmental educators in Greece to introduce new courses of study. Geographer Dieter Böhn looks at the complex issues surrounding EE in the highly industrialized postwar economy of Germany. He analyzes the Bavarian curriculum and emphasizes the importance of recognizing systems interrelationships. In discussing the organizing principles of sustainable development and voluntary simplicity as they are related to the German situation, Michael-Burkhard Piorkowsky describes the impact of household decision-making on environmental quality in a consumer society. He describes four possible paradigms that may help in our analysis.

Reporting on some of the outcomes of the Environmental Education into Initial Teacher Education in Europe (EITTE) project, Chris Oulton and W.A.H. Scott emphasize the role of teacher education in the successful implementation of environmental units in school curricula. Gaoyin Qian and Tiechi Huang address the paradox of a Chinese population policy that puts great pressure on "only children" to focus on classical studies and the resistance to introducing environmental studies into a traditional curriculum even as the country has announced new policies for land use. They further underscore the tension that exists between traditional familial and environmental values in China. Raffaella Semeraro and her colleagues at the University of Padua, Italy, lay out the political constraints involved in implementing EE programs in and out of school in Italy and the need to recognize the complementary goals of Italy's Ministry of Education and Ministry for the Environment. The authors emphasize the importance of conceptual change rather than merely presenting factual information incrementally. They adopt a social constructivist approach, a point taken up later in this volume by Luigina Mortari. Collectively, the chapters in Part I offer a broad view of the issues and options facing the world's environmental educators and policy-makers.

In Part II, the conflicting interests of eco-politics and environmental activism are discussed from several perspectives. Frank Puin discusses the planning of a major international exposition—EXPO 2000 in Hannover, Germany—that has been touted as ushering in a new relationship among "Man, Nature, and Technology." David W. Shapiro challenges us to make a paradigm shift by incorporating a bioregional vision in our policy deliberations. The resistance and disputes over such an approach are currently evident in the management of natural resources both within the United States and across the U.S.-Canadian and U.S.-Mexican borders as well as in the regions of the former Soviet Union. Arline Bronzaft points to the often neglected effects of worldwide noise pollution on health, psychological wellbeing, and children's learning in urban environments. Jack Ahart, a retired member of the U.S. military, addresses the problem of toxic wastes at military sites, a sensitive issue for the United States as well as for the armed forces of other industrialized nations. The question posed—Are environmental issues "vital" in the military sense?—forces us to examine all sources of pollution, however sensitive, if we are to maintain the planet's diverse ecosystems.

Glenn D. Ruhl, an educator and business representative, highlights the role of the business community in environmental education and policy-making in the Canadian province of Alberta. The Alberta project can serve as a model of what might be done if businesses were to genuinely express a commitment to the green—i.e., the environment, not greenbacks. David B. Sachsman explains the vital role of the Third Estate in reporting environmental risk. He points to contributions made by environmental journalists in promoting more accurate reporting to create a better-informed public. Gene McQuillan brings American naturalists and nature writers into the conversation as he examines the impact of the disappearance of wilderness areas on the production of literary works. Nancy E. Wright offers the Luquillo rainforest of Puerto Rico as a case study of changing environmental images that have developed as a consequence of shifting cultural norms seen in political context. Turning attention to Native Americans—the "first peoples" of the North American continent—Isaiah Smithson cautions us against an uncritical acceptance of the "first ecologists" stereotype of Native Americans. We need to better understand their complex and differing relation to the environment and the differing impact on the environment of their varying ritual practices.

In Part III, the problematic relationship of gender and environment is examined by authors from different disciplines. German feminist home economist Barbara Methfessel notes the importance not only of gender roles in households and in society but of the different valuations attached to tasks with environmental impact in everyday life. A sociologist and a home economist, Hiltraud Schmidt-Waldherr reports on a German study of the response of rural women to "environmental" policies made by urban bureaucrats in distant places. Amanda Woods McConney and her co-investigators studied the environmental decision-making of secondary school students and found that, as compared to boys, girls were better able to generate more environmentally sensitive alternatives when making lifestyle decisions. Patricia J. Thompson credits Ellen Swallow-Richards, an early researcher on environmental problems, with having introduced the concept and term *ecology* to American vocabulary in the late 19th century. She fits her contributions into the Hestian/Hermean paradigm. Peter B. Corcoran details the impact of Rachel Carson whose scientific background combined with lyrical writing in *The Sea Around Us* (1956) and *Silent Spring* (1962) alerted the American public of the 1950s and 1960s to the impending environmental crisis brought about by the escalating use of pesticides. Applying an ecofeminist lens to the Woman/Nature connection in Toni Morrison's *Song of Solomon,* Margery Cornwell discloses the tension between the natural/ material that evokes the subtle theme of environmental versus entrepreneurial interests in this important novel. In her study of young children's ideas of Nature, Luigiana Mortari describes their conceptual development along an inclusion/ exclusion schema in an approach compatible with Morrison's in the previous chapter and, as Smithson observed about Native Americans, between the selfish and the cooperative uses of natural resources. The children in the study also express different conceptions of women's and men's relationship to nature. Child psychologist and educator Abigail S. McNamee applies a developmental perspective to explore how an ethic of care is established in early childhood. She links early nurturing to the capacity to care not only for the self-system but for the planetary ecosystem. In his discussion of Diane Ackerman's highly original "take" on nature writing, Gene McQuillan notes that she breaks down the arbitrary division between the natural sciences and the humanities in writing that is at the same time pungent and poignant.

Part IV addresses specific educational strategies necessary for building environmental programs throughout the curriculum as well as the importance of administrative and community support. Nicholas Smith-Sebasto combines a touch of whimsey with serious concern about students' general lack of basic knowledge about the environments they inhabit. Patricia Carlson and Steven McGee describe an innovative use of computer technology to bring students into a "scientific community" of environmental researchers. Martin Stanisstreet and Edward Boyes report on experiences with teachers in recognizing the problems associated with increased vehicular traffic in their own communities. Esther Zager Levine describes a program involving learning-disabled students who participated in an innovative environmental program that allowed them to use computer technology both to access information and to communicate with other environmentally concerned youth. Such programs, I would argue, would be successful with students at all levels of ability and capacity. Kelley M. Shull suggests how the ecological orientation of a writing instructor can approach the act of composition not as a linear sequence of ideas, but as an ecology of ideas with often unpredictable—but potentially satisfying—outcomes.

The challenges faced by an academic administrator in introducing and promoting a new environmental studies curriculum is explained by Rosanne Wille, Provost of Lehman College, whose office must mediate the multiple demands made on an academic administration, especially in a time of shrinking budgets and increased competition for resources among the disciplines. In another "take" on the pragmatics of innovation in higher education, Anne D. Perryman, Director of Media Relations and Publications at Lehman College, presents a summary of successful experiences in planning, organizing, and communicating a college's environmental philosophy that can serve as a model for other institutions as they seek to enlist greater support for their emerging and expanding environmental programs.

This text posed some challenges in reconciling the bibliographical styles and scholarly conventions in different disciplines, using different methodologies and writing in different countries. We have done our best to bring these diverse sources into a useful and "common" style. As further interest in interdisciplinary and international scholarship develops, it will surely find its way to a universally acceptable style and practice in the interdisciplinary, international ecology of an emerging community of environmental scholars, activists, and supporters.

This is the first volume in a projected series *Themes in Environment and Education* which is intended to:

- Promote a global perspective on local, national, and international environmental problems
- Respect the multiple voices that address environmental issues
- Provide a forum for interdisciplinary approaches to environmental education
- Recognize the presence of environmental topics in all academic disciplines
- Share and promote the knowledge in different disciplines necessary to change behavior
- Increase awareness of environmental problems, principles, and policies among professionals and their publics
- Stimulate cooperation among public, private, and nonprofit sectors in examining environmental policies and practices
- Link theory to praxis in environmental policy-making

In sum, the present volume represents an ecology of ideas that can serve as a prototype for conversations we must initiate to promote both interdisciplinary and international themes on environment and education in the 21st century. Such an approach might also include the celebration of nature and environmental themes in cultural studies—the visual and performing arts including drama, music, TV, poetry, painting, film, photography, sculpture. It would also take into account the environmental connections related to food, clothing, and shelter in our everyday lives. Forthcoming titles in the *Themes in Environment and Education* series will not link ecology with a single "environmental discipline." Rather, the scope of the series will allow for examination of ecological/environmental approaches in all the traditional and applied disciplines as well as the impact of environmental issues on politics and the professions. In concert, such approaches will enable those who inhabit different disciplinary domains to cooperate in planning coherent policies that involve private and public decision-making that will collectively benefit our planetary ecology.

Lehman College, CUNY
The Bronx, New York
February 15, 1997

Part I

A Global Perspective

Chapter 1

Beyond Science: Global Imperatives for Environmental Education in the 21st Century

Joy A. Palmer
University of Durham
United Kingdom

Approaches to environmental education (EE) during the past two decades have changed as our understanding of the environment and of the processes of teaching and learning has advanced. Many traditional approaches have been unsuccessful because they were based on narrow perspectives on education, research, communication, and even the nature of the environmental problems we aim to solve. As we move toward the 21st century, our challenge as educators is to develop and engage in more relevant approaches to EE that transcend the limitations of past perspectives. Effective EE will require global efforts to increase its availability. Its success will depend on how well we improve its quality and relevance.

Environmental Education: Defining a Field of Study

The significance of EE has been widely recognized for about 25 years, though views vary as to when the words *environmental* and *education* were first used together. Disinger (1983) writes that the term was used in 1948 at a meeting in Paris of the International Union for the Conservation of Nature and Natural Resources (IUCN). Yet only since the late 1960s has EE been actively debated and promoted globally. A landmark IUCN/UNESCO meeting on "Environmental Education in the School Curriculum," held in Nevada, USA, in 1970, arrived at a definition that was widely adopted around the world:

Environmental education is the process of recognizing values and clarifying concepts in order to develop skills and attitudes necessary to understand and appreciate the interrelatedness among man, his culture, and his biophysical surroundings. (IUCN 1970)

Since that meeting, the translation of global definitions, objectives, and principles into specific policies, programs, and resources at national and community levels has been sought in documents such as *The Belgrade Charter, The Tbilisi Declaration, The World Conservation Strategy, Our Common Future,* and, most recently, *Agenda 21.*

Along with the upsurge of interest in EE, related "educations" have come to the fore: development education, global education, peace education, citizenship education, and human-rights education. All of these have a place on the global agenda and are linked in one way or another to pro-environmental behaviors. They also suggest the tremendous energy that is being devoted to the promotion of planned processes that enable participants to explore and understand the environment and take action to make the world a better place for all of life's forms.

From Environmental Crisis to Behavioral Change

Global acknowledgment of the centrality and importance of the concept of sustainable development is the hallmark of EE in the 1990s—and is sure to make an even greater impact in the 21st century. Developments in the philosophy, policies, and practice of EE leading to this position have been myriad and complex. In general terms, they have transformed the dominant view of the field from one of teaching *about* nature—with "show–and–tell" techniques—in the early 1970s to one of teaching through experiential fieldwork and values education in the 1980s to one of action research and student-led problem-solving fieldwork in the 1990s. These trends are moving us in the right direction. Yet our perspectives are still too narrow, in part because the crises the planet is facing are treated typically as a series of "issues" such as energy, conservation, pollution, and the destruction of nonrenewable resources. This perspective is grounded in the scientific domain but often excludes crucial social, political, and economic concerns. The perspective has also focused on the need to communicate information *about* problems in order to bring about a change in people's behaviors.

I ask: Is this sufficient?

A reliable indication of the leading thinking about EE today may be gained from the nature of the research that is now being undertaken in the field. My reading of major journals and research publications reveals that empirical inquiry is predominantly based on an applied-science model as used in other areas of education, such as science education. Robottom & Hart (1993) expand this argument in their monograph *Research in Environmental Education: Engaging the Debate.* The authors claim:

> Much of the research in environmental education takes the form of ascertaining the congruence between outcomes and assumed goals and seeking empirically (objectively) to derive generalizations (theory) and hence legitimate scientific knowledge. Within this paradigm, the problem of improvement of environmental education continues to be seen as a matter of identifying and controlling variables associated with such goals as responsible environmental behavior. (31-32)

Recent studies in EE are concerned with the identification, prediction, and control of the variables that are believed to be the critical determinants of pro-environmental behavior (Hines, Hungerford, & Tomera 1986; Ramsey & Hungerford 1989).

Another trend in research that aims to inform educational practice focuses on the acquisition and development of environmental knowledge. Stanisstreet & Boyes (this volume), for example, provide a critical assessment of knowledge of scientific ideas underpinning students' understanding of the ozone layer. Their study demonstrates that high school students understand the ozone layer to be a layer of gas around the earth but hold misconceptions and uncertainties about the relationship between the ozone layer and life on earth and about the causes of ozone-layer depletion. The students' naive mental models about relationships between forms of atmospheric pollution and various global environmental problems were not isolated but were part of a consistent framework of confusion and misconception. This study adds to the already well established body of literature on learning and conceptual change in science, complementing the work of researchers such as Roth (1991), Anderson (1987), Anderson & Smith (1987), Carey (1985), and Driver (1983). The data provide a sober reminder that learning that requires changes in students' concepts about scientific issues is very difficult to accomplish. Such studies should compel science educators to develop instructional strategies that will help rectify students' misconceptions and confusion—and untangle the complex web of interrelated environmental issues. If EE is to succeed in its goals, young people leaving high school will need to be able to engage in informed debate about environmental issues with a sound, accurate knowledge of the subject base.

Carlson & McGee (this volume), working in suburban Chicago, show a dramatic increase in high school student interest in environmental issues arising from the creation of an environmental science course that gives students the opportunity to explore issues from a scientific perspective and engage with the scientific community. Their work highlights critical questions pertaining to scientific education: How can students be helped to interpret and draw conclusions from data about the environment? How do they know how much of the data is evidence? How can they create theories about the data?

The examples I have given (and the literature is replete with many others) remind us of the crucial relationship between environmental or ecological education and science. Teachers need to be aware of the following:

- the fact that a sound base of scientific concepts is crucial for environmental understanding
- the importance of accurate scientific knowledge
- the status of misconceptions
- the essential need to develop teaching strategies that may eliminate these misconceptions

The words *science* and *science education* are at the forefront of much of what is presented and published as research in EE. I am not advocating a removal of this link, but rather a shift in emphasis—a broadening of perspectives. My own research has focused on the development of science-related concepts through the study of environmental issues (Palmer 1994). My argument is that EE does not lie exclusively in the field of science education. It is not, and should not be, an "adopted child" of this field—something to be co-opted by science educators alone.

Education *About, In,* and *For* the Environment

Let me emphasize the importance of helping students to gain an ecological perspective—one that comes from acquiring a range of systematic knowledge from the science disciplines of geography, geology, biology, physics, and chemistry. In the traditional EE model, this constitutes education *about* the environment—i.e. the transmission of knowledge that may explicate aspects of the environment or provide conceptual capacity to do this. Environmental education is this and more; it entails a wider interpretation of the environmental crisis that needs to be learned *about,* as well as the all-important facets of education *in* the environment, e.g., experiential field-work aimed at interpreting and appreciating the environment and education *for* the environment aimed at challenging harmful exploitation of aspects of the environment and promoting a caretaker ethic (see McNamee, this volume) on behalf of the environment.

Global Environmental Consciousness through Education

At the level of educational practice, education *about* and *in* the environment has tended to be predominant around the world. Rather rarer attempts to enact forms of education *for* the environment are being promoted in the UK, the USA, and elsewhere with resultant international interest. For example, the Austrian-coordinated "Environment and Schools Initiatives Project" is now in 11 European countries, each with several participating schools and teachers, as well as a national coordinator. Principles guiding the project include "Engagement in Environmental Education" activities characterized by the following:

- personal involvement of students and emotional commitment
- interdisciplinary learning and research
- reflective action to improve environmental conditions, and
- involvement of students in decision making or problem finding, in procedures, and in monitoring their work.

Participating teachers adopted a research perspective with respect to their own teaching and curriculum activities. They were also asked to reflect systematically on their activities in an effort to improve them and to contribute to their own and other people's knowledge of EE. A distinctive feature of this project is its community-based action orientation. The project also requires a key professional role of teacher-as-researcher. Both the teacher's task and the approach to that task are critical. High-quality, effective EE requires as much attention to reflection, evaluation, and planning as any other element of the school curriculum. A successful EE planning model expands on the threefold framework I have mentioned. Tasks should be planned that educate *about* the environment, *for* the environment, and are accomplished *in* the environment. Within this framework, three elements are crucial: personal *experience* in the environment, the development of personal *concern* for the environment, and the taking of personal *action* on behalf of the environment.

Rising to the Challenges of Relevance and Responsibility

What are the global imperatives for EE in the 21st century? First, we need a wider vision of environmental issues. Global environmental problems are numerous and complex. Dramatic increases in the human population and its activities have changed the environment more rapidly in the second half of the 20th century than at any time in our history. The 21st century will have to cope with these changes and reverse them if possible. The environmental crisis should not be treated as a series of resource, pollution, and conservation issues to the exclusion of social, political, economic, and esthetic concerns. Fundamentally, environmental issues are political rather than technical in character. The welcome introduction of the concepts of sustainable use and sustainable development into environmental debate helps give coherence to our understanding of problems. Environmental problems involve concerns about the quality of life and the needs of societies. They involve ethical dilemmas for future generations that go well beyond the scientific domain.

Second, we need to place far greater emphasis on the human dimensions of environmental change. The scale and nature of environment-development interactions are changing drastically as the century draws to a close. We are increasingly faced with extremely complex situations on a wide range of scales. These are driven by—and have an impact on—socioeconomic and political activities, which themselves are in a process of change. Here is an example:

> Overcrowded townships, the air heavy with smog, barren soils scarred by ravines and bereft of vegetation, people and land under threat from toxic waste dumps, polluted rivers and pesticides. South Africa is suffering from decades of environmental mismanagement, aggravated and often institutionalized by apartheid, which forced people to live in rural and urban areas unable to sustain themselves. (Ramphele 1991)

If the human dimensions of environmental issues and environmental change are not thought through, educators may accept the belief that EE content can be identified like other subjects in the curriculum. For example, biology is teaching *about* life in the biosphere, chemistry is teaching *about* chemicals, music is teaching *about* music. But environmental education is not solely teaching *about* the environment. It is about interactions between human life and the environment and the possibilities of overcoming and preventing detrimental environmental impacts in the future. We need a shift in emphasis to focus on the human dimension of environmental change.

Third, we need to accelerate the trend for EE to incorporate action research and community problem solving. Promotion of the ideas of "empowerment" and "capacity building" embrace the significance of students' personal involvement, action, and decision making with respect to socioecological issues in their own communities. Core issues underpinning EE today include equality, social justice, interspecies justice, and intergenerational justice. In other words, EE should move in the direction of more relevant approaches to teaching and learning—approaches that incorporate a wider vision of problems and a methodology compatible with that vision.

Granted, there will never be a 100 percent right solution to community issues. What is "right" depends on one's values and principles (Breiting 1994). Breiting uses the example of teaching about waste and recycling. Tackling this requires moving away from teaching pupils how to participate in a specific system of recycling. Rather, we must focus on their experiences with recycling and on other concrete matters of the actual waste issue relating to people's problems with the use of natural resources. Evaluation of the teaching should be directed toward how well the pupil has learned to understand the background of the environmental issue (e.g., waste/recycling), how well he/she is able to use his/her knowledge to help understand other environmental issues, how well the pupil is able to make up her/his mind concerning what she/he regards as the best solution to a specific environmental issue, and how well she/he is able to act accordingly, both as an individual and in the decision-making mechanisms of the community (Breiting 1994). Let us rise to the challenge of finding appropriate teaching strategies that incorporate action research and community problem solving.

Finally, it is essential that the fundamental importance of EE be recognized at all levels in society—within formal education, nongovernmental organizations (NGOs), and local and national government policy frameworks. Environmental education must be placed at the heart of policy and curriculum-development processes. In the 21st century, educators need to do the following:

- engage in both a vision and a practice encompassing all aspects of sustainable living within life-support systems and processes

- devise policy frameworks, research, and curriculum-development processes that will reflect this vision
- be at the forefront of an empowering process through which all of us will be enabled to appreciate—and engage with—the true complexity of environmental interactions of the time

References

Anderson, C.W. 1987. "Strategic Teaching in Science." In B.F. Jones et al., eds. *Strategic Teaching and Learning: Cognitive Instruction in the Content Areas.* Association for Supervision and Curriculum Development, Alexandria, VA & North Central Regional Educational Laboratory, Elmhurst, IL, 73-91.

Anderson, C.W. & E.L. Smith. 1987. "Teaching in Science." In V. Koelhler, ed. *The Educator's Handbook: A Research Perspective.* New York: Longman.

Breiting, S. 1994. "Towards a New Concept of Environmental Education." Paper presented at the Conference on Exchange of Promising Experiences in Environmental Education in Great Britain and the Nordic Countries. Karlslunde, Denmark (November).

Carey, S. 1985. *Conceptual Change in Childhood.* Cambridge, MA: MIT Press.

Disinger, J. 1983. "Environmental Education's Definitional Problem," *ERIC/ SMEAC Information Bulletin* 2. Columbus, ERIC/SMEAC.

Driver, R. 1983. *The Pupil as Scientist.* Milton Keynes: Open University Press.

Driver, R., E. Guesne, & A. Tiberghien, eds. 1983. *Children's Ideas in Science.* Milton Keynes: Open University Press.

Hines, J.M., J.R. Hungerford & A. Tomera. 1986. "Analysis and Synthesis of Research on Responsible Environmental Behaviour: A Meta-Analysis." *Journal of Environmental Education* 18: 2,1-8.

IUCN. 1970. *International Working Meeting on Environmental Education in the School Curriculum: Final Report,* September 1970, IUCN, New York.

IUCN, UNEP, WWF. 1980. *The World Conservation Strategy.* New York.

Palmer, J.A. 1994. "Acquisition of Environmental Subject Knowledge in Pre-School Children: An International Study," *Children's Environments* 11:3, 204-11.

Ramphele, M. 1991. *Restoring the Land.* London: The Panos Institute.

Ramsey, J. & H.R. Hungerford. 1989. "The Effects of Issue Investigation and Action Training on Environmental Behaviour in 7th Grade Students." *Journal of Environmental Education 20*:4, 29-34.

Robottom, I. & P. Hart. 1993. *Research in Environmental Education: Engaging the Debate,* Australia: Deakin University.

Roth, K.J. 1991. "Reading Science Texts for Conceptual Change." In Carol M. Santa & Donna E. Alvermann (below), 48-63.

RSPB. 1994. *Biodiversity Challenge: An Agenda for Conservation in the UK.* Sandy: UK.

Santa, Carol M. & Donna E. Alvermann, eds. 1991. *Science Learning: Processes and Applications.* Newark, DE: International Reading Association.

UNCED. 1992 *Agenda 21,* United Nations Conference on Environment and Development (The Earth Summit).

UNESCO. 1978. *Final Report: Intergovernmental Conference on Environmental Education* organized by UNESCO in cooperation with UNEP, Tbilisi, USSR.

World Commission on Environment and Development. 1978. *Our Common Future* (The Brundtland Report). London: Oxford University Press.

Chapter 2

Human Ecology: Environmental Education in Greece

Jacqueline Gravanis
Harokopio University
Athens, Greece

In Greek mythology, the image of the environment revealed a divine nature; gods, demigods, and human beings formed a unity on Mother, Earth which also included woodlands, rivers, and animals. As the divine aspect of nature, the gods gave priority to preservation of the environment (Vallera 1990, 49). The environment became holy and inviolable.

Ancient Greeks of the Classical period were not merely surrounded by the environment but literally possessed by it (Bouratinos 1990, 60-64). Greek philosophers and historians writing about "nature" defined basic environmental principles. Hippocrates saw "the air, the water, and the various locations" as apart from the people who were influenced by the environment; thus, people's health was the result of those conditions. Xenophon, in *Oekonomikos,* introduces ideas of productivity and management of the environment. Aristotle valued the connection between specific needs and consumer goods so that man could transcend his avarice (Vallera 1990, 50).

During the 19th century, the study of the physical environment was more systematically organized; Haeckel (1834-1919) was the first to use the term *ecology,* and in 1866, he included the "Study of Areas" in *Natural Economy* (Stram 1986, 13). Technological developments in the 20th century have enabled humankind to acquire enormous power and consequently cause great ecological damage to the environment. This has been "a century of destruction and decomposition in the name of prosperity and evolution... [T]he brutal and inconsiderate intervention to the earth, to the air, to the plants, is being done for the sake of progress" (Polimilis 1992, 10). This degradation has led to the need for action to protect and preserve the environment. The "preservation movement" emphasizes the problematic relationship between human beings and the environment (Flogaitis 1993, 70). This is the context for the introduction into educational systems of what is now called EE.

The Rise of Environmental Education

Environmental education as we know it began in the 1960s in response to concerns in the international community about environmental conditions and crises. Methods and concepts were developed at national meetings in the 1970s under the supervision of organizations such as OHE, UNESCO, the European Committee, and the European Community. The aim of EE was to promote the protection and conservation of nature, mainly through the formation of programs of "study of the environment" (Lahiry et al. 1988, 4). At the first international meeting, organized in 1970 by the International Union for Conservation of Nature and Natural Resources (IUCN) in Carson City, Nevada, EE was defined as follows:

> the process of identifying the values and clarifying the particular concepts leading to the development of skills and the formation of attitudes necessary for understanding the relationship between man [sic] and his civilization as well as his biophysical environment. (UNESCO-UNEP 1985, 9)

Subsequently, UNESCO and UNEP organized an EE program that included research studies, skills seminars, and a variety of experimental projects. The international program published extensive related materials and organized meetings where EE issues were discussed and analyzed. These activities led to major international conferences, such as those in Belgrade (1975), Tbilisi (1977), and Moscow (1987).

The OHE conference on human ecology in Stockholm (1972) determined the direction of EE with its "Declaration of Human Ecology," in which 26 principles were cited. The aim of these principles was to "inspire and lead the nations of the world so as to conserve and improve the environment" (UN 1973). This Declaration proposed a plan of action based on international cooperation to deal with environmental problems. The philosophy, aims, and principles of EE formed during the Tbilisi conference were later adopted by the conference in Moscow.

The Greek Experience in Environmental Education

Greece has a complicated ecosystem of particular interest internationally. This chapter will discuss the Greek environmental problem, its effects, and some of the actions that have taken place. As in both developed and developing countries, the rate of degradation of the natural environment in Greece has been alarmingly high, especially in recent years. The degradation has resulted from (1) forest fires, (2) overexpansion of stock breeding and agriculture, (3) excessive use of pesticides, (4) the establishment of industrial areas without following the appropriate regulations, (5) the lack of organized planning for residential land in and around cities, (6) overpopulation in the cities of Athens and Salonika (where about 50 percent of the Greek population is concentrated), and (7) the abandonment of agricultural areas and of remote islands. These consequences are due to scientific and technological developments that have enabled humans to "control" nature. This has resulted in improved living conditions, on the one hand, and a seriously damaged environment, on the other. Since 1975, the Greek government has systematically recorded environmental problems and dangers. Public awareness and sensitization has not been adequately developed, however, and a greater effort must be made through the educational system.

As O'Riordan (1976, 314-15) notes, EE can be nothing short of the training of citizens to understand their relationship with Nature. That is, EE's aim is not the domination of human beings over the environment, but the harmonious coexistence of people and other life forms on the planet. Through EE, a new way of thinking about cultural values and quality of life is being formed. Environmental education has been an official part of education programs in Greek schools since 1991, but it was introduced in 1977 (Spiropoulos 1977, 36-39) and has developed in three phases.

Phase I: 1977-1982

In the first phase, the Educational Institute, in collaboration with the Secretariat of the National Council for Environment and City Design, undertook the promotion of EE. The outcome of this collaboration included the following events:

- development of a bibliographical source of information about EE

- training of 20 Greek science teachers at environmental centers in France for three 45-day periods
- organization in 1980 of teacher-training seminars on EE by instructors from the European Council
- introduction of basic ecology principles in school textbooks
- teaching of a new subject, environmental studies, in grades 1 through 4 of elementary schools
- participation of Greek schools in the primary and secondary education network of the European Community EE program (1981-86)
- introduction of a pilot EE program, entitled "Protection of Coasts," in four Greek junior high schools (gymnasia) (1981-86) by the European Council (Gardeli 1984, 57-72)
- issuance and distribution of "Man and the Environment" posters to junior high school students by the Ministry of Education (Glavas 1992, 29)

During the first phase, only a small number of primary and secondary students and teachers participated. These programs were offered within the framework of "cultural activities" in the junior high school curricula (Spiropoulos 1977, 36-9).

Phase II: 1983-1989

The second phase included the following events:

- A team of EE-trained teachers from the Educational Institute undertook responsibility for (1) promoting EE programs, (2) training teachers, (3) troubleshooting the new EE programs, (4) evaluating the programs, and (5) making suggestions.
- An office responsible for promoting EE was opened under the Department of Secondary Education at the Ministry of National Education. Since 1987, the Department of Primary and Secondary Educational Studies of the Ministry of Education has taken responsibility for promoting EE at all levels of primary and secondary education.
- About 1,500 teachers throughout Greece were trained at 20 sessions of from two to ten days each (Glavas 1992, 30). At the same time, EE was also introduced in seminars for newly appointed teachers, as well as in training schools for teachers of primary and secondary education.
- A seminar was held in Athens, in collaboration with other countries in the network (1986), with its main objective that of drawing conclusions.

- A brochure from the Ministry of National Education on "School and Environment," with basic information on EE, was distributed (Ministry of Education OEDB 1989).
- Connections were established with international governmental organizations and nongovernmental organizations (NGOs) regarding EE matters (Council of Europe, European Community, UNESCO).

During this phase, some 600 programs were introduced in secondary education, and another 1,000 in primary education. Yet even after all these efforts, many teachers still were unable to recognize the term EE and thus ignored its meaning and value (Glavas 1992, 30).

Phase III: 1990 to the Present

The third phase was characterized by the following events:

- Environmental education was recognized as part of the primary and secondary school curricula with the aim for students to (1) realize the relationship of humans to their natural and social environment, (2) become sensitized to the problems concerning the environment, and (3) take action by participating in international efforts to deal with these problems. In addition, the Ministry of Education opened EE centers throughout Greece and appointed in each region an "accountable employee" responsible for EE programs.
- EE was promoted regionally, with activities coordinated in local schools by the "accountable employees." UNESCO-sponsored and other employee-training seminars were held.
- The first national center was established for the promotion of EE and the implementation of programs sponsored by the Ministry of Education, in collaboration with other ministries, local authorities, and NGOs, such as the Hellenic Society of Environmental Education and Information, the Hellenic Society for the Protection of Natural and Cultural Inheritance, the Association for the Protection of Sea Turtles, the Hellenic Bird Society, the Hellenic Union for the Protection of the Sea Environment, the Hellenic Union of Aluminum, and the World Wildlife Fund for Nature.

The third phase of work in primary and secondary education can be considered a landmark in the development of EE in Greece. Apart from its introduction to the curricula, EE specifies methods of application as well as the financing of these mandates. New horizons have also opened for EE. In the first year after the appointment of the "accountable employees," more than 1,500 EE programs were carried out in primary and secondary schools (although these have been on an optional basis, offered mainly as extracurricular activities). According to Greek law, EE programs can be introduced in the regular weekly curricula under the following conditions:

- When the particular class or section participating in the EE program has a 20-hour-per-week program, two extra hours of EE teaching are added, totaling 30 hours weekly.
- When the weekly program has 29 or 30 hours, two extra hours of EE teaching are added, totaling 31 or 32 hours a week. Teachers participating in the program may choose to teach EE, using some of their teaching time, on the condition that their subject is directly related to EE programs.

The EE-related subject areas accepted by the Ministry of Education include the following:

- ecological elements that have undergone deterioration: (1) ecology (soil, earth, subsoil, topography, sources of energy, water, climate, landscape, ecology systems, subsystems, flora, fauna, natural habitat, woodlands); (2) elements of human ecology (ancient towns, traditional residential monuments, human-made creations, customs, microsocieties, and arts and crafts); and (3) elements of physio-human ecology (units of nature and civilization)
- sources and causes of ecological deterioration: (1) areas of economic activity (agriculture, fishing, hunting, herb collecting, tourism, mines, factories, transportation means, draining systems, technical projects, and generation of energy); (2) development of human activity (economic, technological, and sociocultural); (3) wars; and (4) catastrophes, biochemical pollution in the atmosphere, water, or soil; thermic; biogenetic noise (see Bronzaft, this volume); radio energy; and pesticides)

Until 1984, elements of EE in higher education were provided by the Department of Forestry and Natural Environment at the University of Salonika and by the Departments of Biology, Physics, Chemistry, Economics, and Agronomy at universities, as well as by certain departments in the polytechnic institutes. The above-mentioned departments examined the environment from a scientific point of view according to their disciplines. Subjects concerning EE are also taught in Departments of Education through subjects such as ecology and geography. At the University of the Aegean, founded in 1987, the Department of Environment offers environmental studies focusing on the content and management of ecosystems, with an emphasis on problem solving.

The newly established Center for Environmental Education, which operates under the Department of Philosophy-Education-Psychology at the University of Athens, has as its goal the best possible EE for teachers.

Conclusions

Environmental education in Greece is a relatively recent innovation with high goals, but a number of problems affect its complete incorporation in educational offerings at all levels. As noted earlier, public awareness has not been adequately developed, and a greater effort to achieve this is needed through the educational system. Despite recent commitments to public sensitization, the present situation may still be characterized as inadequate because of: (1) insufficient school equipment, (2) ineffective use of free time, (3) negative use of youth dynamics concerning environmental problems, (4) inadequate scientific research in EE, and (5) the lack of continuous and well-organized training for primary and secondary EE teachers.

Because it is important for all citizens, as well as students, to be well informed on environmental issues, an ongoing effort is under way to reach a wider public through lectures in local training centers and seminars organized by the Ministry of Education and EE advisers. In contrast to its first years in Greece, the enthusiasm with which both youngsters and teachers now welcome EE is evident. This is encouraging because, despite its problems, EE will become an important initiative in Greek schools. The objectives and goals of EE are known; the same applies to teaching methods. It is imperative to introduce effective educational, as well as environmental, policies so that EE can play its proper problem-solving role. Indeed, EE concerns everyone—especially the Greeks who, from their earliest history, considered themselves part of the "greater world." Is it possible that the sacred and inviolable environment of Greek mythology is the lost paradise that EE has sought to recover?

References

Alexopoulou, I. 1992. "The policy of EU (European Union) for topics on environmental education after 1988." In *Text for environmental education*. Athens: Ministry of Education, UNESCO, 23-33.

Bouratinos, A. 1990. "Sacred Environment in Ancient Greece." *The New Ecology*, 60-64, 70-71.

Bourodemos, E.L. 1990. "Environment and Development in Greece." Athens: Axiotels.

Buzzati-Traverso, A. 1977. "Some Thoughts on the Philosophy of Environmental Education." In UNESCO ed., *Trends of Environmental Education* (13-23), Paris: UNESCO.

Caduto, M. 1985. "A Guide on Environmental Values of Education." EE Series 13 (UNESCO-UNEP-IEEP), Paris: UNESCO.

Christias, I. 1985. "The Study of Environment: Topics for Teaching." Athens.

FEK 175, v. B', March 19, 1993.

FEK 306, v. B', April 29, 1993.

FEK Government Gazette: 101, v. A', N. 1892. July 31, 1990. G2/3026/Aug. 27, 1990, Ministry of Education.

FEK 629, v. B', Oct. 23, 1992.

Flogaitis, E. & I. Alexopoulou. 1991. "Environmental Education in Greece." *European Journal of Education 26*: 4, 339-45.

Flogaitis, E. 1993. *Environmental Education*. Athens, GA: Athens University Press.

Gail, P.A. 1976. "An Environmental Guide for Teachers." Ohio Institute for Environmental Education.

Gardeli, S. 1986. "Education et Environment en Grêce, une Experience Pilote. L'école et Communauté." Paris:Ministère de l'Education Nationale (Direction des écoles) et Ministère de l'environment.

Glavas, S. 1992. "The Ministry of Education and Environmental Education." In *Report of the Fifth Educational Conference: Environmental Education*. Athens, GA: Athens College, 28-32.

Goulandris, A. 1992. "Meeting with the Present and the Future." *Eleftherotypia* (June 6), 9.

Kasimanis, K. 1991. "The Ecological Catastrophe and the Refutation of Humanism." *Education* (Greece) 24, 116-29.

Lahiry, D., S. Sinha, J. S. Gill, U. Mallik, & A.K. Mashira. 1988. "Environmental Education: A Process for Preservice Teacher Training Curriculum Development." P.R. Simpson, H.R Hungerford, & T.L Volk, eds. EE Series 26 (UNESCO-UNEP-IEEP), Paris: UNESCO.

Markandonis, I.S. 1973. "Free Time During the Age of Technology," Athens.

Misios, X. 1992. "The Relationship with Nature is Maternal." *Eleftherotypia* (June 6), 13.

O'Riordan,T. 1976. *Environmentalism.* London: Pion.

Papademetriou, B. 1989. "Problems Regarding Environmental Education." *Contemporary Education 44,* 57-62.

Papademetriou, B. UNESCO-UNEP. 1977. "Intergovernmental Conference on Environmental Education." *Final Report, Tbilisi* (USSR), 14-26.

Pepper, D. 1984. *The Roots of Modern Environmentalism.* London: Croom Helm.

Polimilis, A. 1992. "Coffeehouse Greece," *Eleftherotypia* (June 3), 10.

Spiropoulos, H. 1986. "Environmental Education in Greece. A First Estimate." *Neoelliniki Paedia 7,* 72-84.

Spiropoulos, H. 1977. "Topics on Environmental Education." In *Cultural and humanistic activities in the Junior High School.* Athens, 36-39.

Stram, E. 1986. "The ABC Book of Ecology," Vol. A. Athens: Eolos.

UNESCO-UNEP, eds. 1985. *A Comparative Survey of the Incorporation of Environmental Education into School Curricula.* EE Series 17 (UNESCO-UNEP-IEEP). Paris: UNESCO.

United Nations. 1973. *Report of the UN Conference on the Human Environment,* Stockholm (June 5-16,1972) New York: UN.

Vallera, E. & M. Korma. 1990. "Environment and Antiquity," *Archeology 35,* 48-52.

Veikos, T. 1991. *Nature and Society: From Thales to Socrates.* Athens: Smily.

Chapter 3

Environmental Education in Germany: An Overview

Dieter L. Böhn
Institut für Geographie der Universität Würzburg
Würzburg, Germany

Germany claims to be one of the most environmentally conscious countries in the world. Nearly every German household separates its waste into at least four bins—one for paper, one for glass, one for bio-waste, and one for all the rest. The new name for garbage is *wertstoffe* ("valuables"). Recycling valuable materials is an important concept in environmental protection. Nearly every article one buys in Germany has a green dot that means it is recyclable, and consumers pay indirectly for German recycling. But this is only part of the picture. Another part is that many Germans do not really want to change their lifestyles. They want to drive their cars as often as possible—and no politician dares to argue for a speed limit. Is the careful management of resources a loss for the economy? Every day advertisements urge us to buy something new, although we could use our cars or clothing for longer periods. There's a big difference between making a minor change of habit and a real change in lifestyle. This becomes a problem for society and for education. What are the aims of EE in Germany, and how is it taught?

The Environment: One Word with Different Definitions

There are two main definitions of *environment* in the German curriculum. The first is "nature." Therefore, EE should teach that human intervention endangers nature, and if we are to protect nature, we must minimize human intervention. However, this approach is widely viewed as "negative." The second definition of environment is "everything around us"—human-made landscapes, buildings, and roads, as well as mountains and waterfalls. Hence, EE should also teach us to protect human-made resources as well as nature. The preferred definition of the environment in EE depends on the subject taught. Biology, physics, and chemistry classes emphasize structures and processes within nature; geography, social studies, and history deal mostly with interrelations between human beings and nature.

An Important Issue, Not an Independent Subject

In 1980, the environment became an important topic in a nationwide public debate over what to include in the German curriculum. A consensus was arrived at—not an easy feat in a country where each state is proud of its own system of education. According to a joint declaration from all of the State Departments of Education:

> For each of us and humankind collectively our relationship to the environment has become a matter of survival. Therefore, it is one of the schools' duties to make students aware of the environment, to make them willing to support responsible management of the environment, and to promote environmentally responsible behavior after school hours.

The Standing Conference of the Ministers of Education and Cultural Affairs decided that there would be no new subject called EE, but rather that EE would be integrated into nearly every existing subject, especially in biology and geography. The rationale for not creating EE as a new school subject includes the following points:

• Teachers in Germany are certified in subject areas, such as biology and geography. There is no certification in EE at teacher training colleges and universities.

• Many other issues, such as peace education, human rights, health education, information technology studies, one-world education, media education, and Europe as a spatial, historic, economic, and social entity, are considered as important as EE. Each of these issues generates topics relevant to the student's present and future life. Yet, compelling as each issue may be, one cannot eliminate traditional subjects such as geography, history, English, or mathematics from the curriculum. At the moment, Germany has about 50 different curricula in geography alone, so there are hundreds of curricula for all subjects in the 16 German states. Environmental education has no discrete curriculum, but it offers numerous issues and topics of interest to other subject areas.

Guidelines for Environmental Education

Let's take the example of Bavarian curriculum guidelines to illustrate the general aims of EE in Germany:

• to promote a loving attitude toward nature and a respect for creation
• to understand the multiple and mutual relationships among nature, humanity, and the environment (in considering these relations, one should understand the responsibility of each person and of the entire community for the wellbeing of the environment)
• to make one willing and ready to take ecologically necessary steps that go beyond self-interest

Within the EE guidelines topics include:

• variety, individuality, and beauty of nature: diversity of species; nature as a place of experience and knowledge; nature as a theme of poetry, art, and music
• the importance and history of the cultural landscape—civilization as a factor of environment: the agricultural landscape, its history and its change, villages and towns as living areas and humanity's home, regional development, preservation of landmarks
• basic rules of ecology: natural and artificial ecosystems, urban ecology, limits of ecosystems
• relationships among the environment, society, state, and the economy: (1) demands made by society and their effects on the environment (e.g., prosperity, full employment, social security, economic growth, freedom of mobility, leisure time, sports); (2) environmental problems of different economic sectors (e.g., energy supply, traffic, industry, architecture, agriculture, tourism, waste economies, environmental protection as a state duty, competition with other duties of the state, international relations and the environment, environmental protection as an international responsibility, implications for industrial states and the Third World); and (3) important environmental principles (environmental protection laws, technical applications for environmental protection)
• the type and magnitude of environmental burdens: local and global impact of human activities on air, water, soil, climate, flora and fauna, landscape, human groups, and culture; single influences; added influences; relationships among influences

• environmental problems as a challenge for science and technology: possibilities and limits of technology for environmental protection; energy conservation and environmental protection; energy supply as a problem of research and development; conservation of resources, minimizing waste, recycling; avoiding environmental problems by long-range planning and research and by using highly developed techniques in research, industry, and the home

• the relationship between human beings and the environment as an ethical challenge: the relationship of values and habits to environmental problems; anthropological insights for forming an ecological awareness; cultural evolution (e.g., curiosity, research, values, the human being as an individual and as a part of a community with a common history); nature as a source of values and faith; ambivalence of scientific and technical progress; guidelines for behavior responsive to the demands of the environment (respect for creation, intrinsic value of the natural environment, appreciation, rating of values, responsibility, readiness for restraint, and doing without)

• personal lifestyle and environment: patterns of consumption; managing everyday life; food, hygiene, and health; work habits; ways of spending leisure time, e.g., sports; transportation choices; participation in duties of all citizens

• history of environmental problems: causes and consequences of population growth; environmental behavior in the course of history (e.g., neglect of ecological principles, ideologies of growth)

Integrating Aims in Subjects: An Official Program

Since EE is not a separate subject, the problem is how to divide the various topics among the different subjects. Again, I will illustrate with examples from Bavarian curriculum guidelines:

Topic 1: diversity, individuality, and natural beauty—nature as a place of experience and knowledge:

• subjects in grades 1-4—basic science, social education, sports
• subjects in grades 5-10—biology, geography, sports, religious education/ethics
• subjects in grades 11-13—biology, sports, geography

Topic 4: relationships among the environment, society, state, and the economy—international relations and the environment; environmental protection as a duty of the international community; industrial countries in the Third World:

- subjects in grades 1-4—basic science, social education
- subjects in grades 5-10—geography, social science, biology, chemistry, economics
- subjects in grades 11-13—geography, biology, chemistry, economics

It is evident that biology and geography are the most important subjects in EE. However, there is no coherent relationship among topics, grades, and subjects. For example, in Topic 1 ("nature as a place of experience"); why is ethics taught in grades 5-10 but not in grades 11-13? There seem to be arbitrary decisions in the assignments of topics.

Let's look at two examples of EE in a German curriculum which illustrate two methods of teaching EE. In the first example, from Realschule, Bavaria, EE is covered in geography:

- grade 7, Topic 3: natural and cultural characteristics and problems of Southern Europe—the ecological burden on the Mediterranean Sea—integrated topics for geography, history, and EE
- grade 9, Topic 2: Russia and other former states of the USSR—important agricultural and industrial areas, factors of location and development in selected areas, Moscow metropolitan area, Ukraine—integrated topics for geography and EE

It is clear that EE is merely an appendage to the main concerns of regional geography, physical geography, and human geography. Thus, German students will not become aware that environmental protection is involved in a system. They will not realize the system of related elements that influence nature; nor will they realize the system of interrelations between human beings and nature; nor will they recognize the various values that motivate different activities. Environmental education is a patchwork with topics on Italy, on the USA, on Germany. German students study a patch here, a patch there, but with no integration. The presentation of EE in other subjects is similar to that in geography.

The second example, from Integrierte Gesamtschule in Lower Saxony, is EE in social studies. Lower Saxony takes a different approach to EE, with "key problems" determining the topics. This may be explained by the key problem "environmental preservation." Comprehensive topics include the following:

- Are we drowning in garbage?
- How much "quiet" does a human being need?
- There is something in the air—poison enters with every breath.
- If all wells (do not) flow, will we need drinking water tomorrow?
- Are Tuareg nomads keepers or destroyers of the environment?
- Is our home town no place or a place for children?
- There is a race for human survival, one of the problems of population growth.
- Does famine in the Sahel make the North Sea and the Baltic Sea a tourist paradise or a sewer?
- What are the hazards of unpredictable nature?

Environmental education is important in this curriculum; it is a topic in its own right. This may be the way EE will be taught in other German states in the future. I would argue that it is important to explain to students how endangered our planet is, but why should we emphasize catastrophes above everything else?

Environmental Education: The Reality in Daily School Life

If one considers only the guidelines and the curricula, EE in Germany seems exemplary. The reality is quite different. For example, some students are unwilling to participate; they enjoy consumerism and see EE as an attempt to restrict their freedom to choose. Some students are bored: nearly every day they are exposed to environmental issues, in school and on television. And because the German society is a consumer society, the EE taught in school is very effectively "countered" outside the school with messages from advertisers, peers, and even the family. While EE calls for voluntary restrictions in our individual lives to benefit the environment, society seems to dictate consumption for the sake of the economy and for social status (see Piorkowsky, this volume).

There are positive experiences as well. Younger students are more willing to work for a better environment; they collect garbage, reduce waste, and prefer recyclable materials. Sometimes students are successful in changing aspects of their own environment; they use recyclable paper, promote solar energy in their schools, and use reusable ceramic cups and plates instead of those made out of paper or plastic.

Environmental education's effectiveness depends largely on the teacher. It is the teacher's personal commitment that convinces students. Thus, EE in school should begin with teachers. Up to now, because most teachers have no special training in EE, only those who are truly interested have been able to teach it successfully. One of the major duties of universities and teacher-training colleges is to prepare students and teachers for effective EE (see Oulton & Scott, this volume). There are good examples in some German universities, but there is still much to be done.

Environmental Education in the Future: Problems and Issues

The Scientific Issue

Many studies have shown that the global temperature is increasing; one name for this phenomenon is the *greenhouse effect*. Scientists have argued that global warming is caused by human beings—through high levels of dangerous emissions from industry, motor traffic, and agriculture. Scientists have also argued that the climate has changed globally many times. During the Roman Empire (A.D. 100-400) and the Middle Ages (A.D. 1000-1300), temperatures were higher than today's global average of 15 degrees Celsius/59 degrees Fahrenheit. During the "Little Ice Age" (A.D. 1400-1830), the average global temperature was only 14 degrees Celsius/57 degrees Fahrenheit. Since about 1850, the global temperature has risen approximately 0.5 degrees Celsius/1 degree Fahrenheit, while the amount of carbon dioxide has risen from 280 ppmv in 1800 to 360 ppmv in 1995.

This phenomenon raises two important questions: (1) What is the human impact on climatic change? If there is no major influence, there is no need to change our behavior. (2) How dangerous is the warmer climate? Some scientists warn of migrations of millions from South to North and furious wars of the desertized South against the still humid North. They also warn of floods and droughts in the North, as well as serious reductions of agricultural productivity and a deterioration of the entire environment.

The Political Issue

At the Berlin Conference on Climate Change (1995), the importance of political, as well as scientific, issues became clear. The most industrialized countries are the most responsible for global environmental pollution, but those countries have been generally unwilling to reduce their output. Above all, the United States has not issued precise data or committed itself to any figure for reduction. Germany has offered to reduce its carbon dioxide emissions by 25 percent, compared with 1987 levels, over the next ten years. Until now, German reductions have resulted mainly from economic problems in East Germany. To make matters worse, in Germany and in other heavily industrialized countries around the world, popularly elected politicians assume that voters are unwilling to change their lifestyles in order to protect the environment or even to avoid changes in the climate.

Newly industrialized countries have suggested that the climate issue is a trick by industrial superpowers to limit the emerging countries' economic growth. Some argue that this is another way for imperialism to prevent their countries from reaching the status and wealth of the highly industrialized world powers. Their position is that they must focus their efforts on helping their people survive now rather than trying to "save the world" in the future. They say: Let the most industrialized countries take care of environmental protection.

The Social Issue

There are two conflicting social trends with regard to environmental protection. The first is that even though people may accept the idea that the environment needs to be protected, they are unwilling to reduce their personal comfort or convenience if that's what it takes. New arguments add to the confusion because they seem to offer proof of opposing positions: (1) that the environment is already ruined, and (2) that the environment cannot be ruined. Of course, both positions have the same result: they eliminate the need for environmental protection!

The second social trend is that environmental protection has become widely accepted as an economic benefit. Many products in Germany are commended as recyclable, as "environment-friendly." In Germany, the prefix *bio* helps to promote the sale of a great many foods and food products. Environmental technology is one of the areas in which Germany is a leader. Economic arguments have become the major influence on changing social values.

Accepting the Problems, Continuing the Efforts

In summary, German educators have positive ideas about EE because it has become a main issue in curricula. Many teachers are working effectively with students to create a new system of values in which the environment holds greater importance, and this should lead to a greater willingness to institute lifestyle changes. An important force working against effective EE is not the economy or politics, but human apathy.

After all, EE is not easy. When children express fears that inescapable catastrophes will occur, it is the teacher's duty to give them a sense of purpose and confidence in the future. We need to change our lifestyles, but that does not mean we have to become poor. We can also use technology to improve the environment. Scientists will inevitably have different opinions about our future, but we cannot wait until nature shows us who is right. We have to consider negative influences, but we still have opportunities to change. Since environmental protection is a relatively new value in society—and since environmental technology has become a promising option—we may have the chance to see EE succeed.

References

Because the numerous publications about EE in Germany are almost exclusively in German, only the guidelines and curricula cited are listed here.

Bayerisches Staatsministerium für Unterricht und Kultus. 1990. *Umwelterziehung an Bayerischen Schulen.*

Bayerisches Staatsministerium für Unterricht und Kultus. 1993. *Lehrplan für die Realschule.*

Kultusministerium Niedersachsen. 1994. *Lehrplan-Entwurf Integrierte Gesamtschule.*

Standige Konferenz der Kultusminister der Länder der Bundesrepublik Deutschland. 1980. *Umwelt und Unterricht.*

Chapter 4

Paradigms of Sustainable Consumption Styles: Approaches and Experiences in Germany

Michael-Burkhard Piorkowsky
Rheinische Friedrich-Wilhelms Universität
Bonn, Germany

Sustainable development is a grand paradigm of environmental science and a main goal of national and international environmental policy. In its broadest outlines, the concept of sustainability has been widely accepted and endorsed. However, as stated by the World Resources Institute (1992, 2), the process of defining the concept in operational terms has not ended. Translating sustainable development into practical goals, programs, and policies around which nations might coalesce has proven even harder to accomplish—in part because countries face widely varying circumstances. For affluent societies, sustainable development means steady reductions in wasteful levels of consumption of energy and other natural resources through improvements in efficiency and changes in lifestyle. Thus, sustainable development implies sustainable consumption.

In market economies with parliamentary political systems such as Germany, households play a key role in the economic process—as consumers of household commodities, as workers, and as voters. Practically speaking, households are the only institution able to reinvent and realize lifestyles that are compatible with the concept of sustainability. However, for day-to-day decision making, abstract concepts (such as the paradigm of sustainability) are less important than strongly held beliefs—for example, in the idea of natural balances within ecosystems or in the importance of living in harmony with other species. According to such beliefs and aspirations, four main approaches to lifestyles and related consumer behavior can be distinguished that (directly or indirectly, in whole or in part) are beneficial to sustainability. These approaches are voluntary simplicity; respect for life on earth; natural food; and new technology.

These four approaches to more sustainable development can be distinguished on theoretical grounds as well as on the basis of empirical data. This classification is to be demonstrated as follows:

- An outline of the interrelations of the economic system and the ecosystem is offered, with special attention focused on environmentally conscious consumer behavior.
- Basic paradigms of sustainability and related types of sustainable consumption styles are discussed.
- Selected data from various empirical investigations of consumer behavior and case studies of sustainable consumption styles are presented.

The Economic System, Private Households, and the Ecosystem

The purpose of economic activities is commonly observed in either one or two functions: to overcome scarcity of resources and to meet the needs of human beings. This can be understood as a consequence of the perception that nature does not provide enough goods, and/or that the goods it provides are not always desired. From this point of view, the fact that economic activity is also a dissipative process has been neglected. The economy is based on a complex, self-organizing system of information-generating and material-processing, driven by a flow of free energy. The system necessarily involves both the transformation of matter and energy and the output of waste, which consists of degraded high-entropy materials that include harmful substances and dissipative energy (Boulding 1968; Georgescu-Roegen 1970). According to the "Law of Entropy," the Second Law of Thermodynamics, this process of degradation is irreversible.

Private households can be seen as socioeconomic processing units and basic institutions of the economy. Their economic function is to organize the first and final activities in production and consumption: to satisfy the needs of household members. Moreover, the household sector is a basic subsystem of society. Household production requires market goods, public services, and natural resources to produce consumable commodities. Consumption leads to enjoyment and to reproduction—through growth and in compensation for energy loss. Negative side effects or diseconomies, which cannot be avoided, are waste outputs in a more or less static condition.

The economic activity of a single household will not normally lead to an overloading of the natural environment. But the household sector as a whole—that is, all of the private households in a national economy—has a great impact on the ecosystem. In market economies, the stabilizing controls are based on labor input and consumer demand. In societies with parliamentary political systems, adult household members in effect vote for public goods. Thus, private households play a significant role in affecting both the state of the economy and the natural environment.

In affluent countries, such as Germany, environmental damage is less a result of high population or population growth than of per capita production and consumption. Thus, environmental protection has become a political issue that is widely discussed in the media. No wonder the level of awareness of the population is high in Germany. Several investigations of environmentally conscious consumer behavior show that virtually every household is doing something to reduce environmental damage. Examples of this are (1) refusing to consume certain goods, (2) shifting demand to products less dangerous to the environment, and (3) sorting and disposing of household refuse (Piorkowsky 1988). However, these are modest efforts, and the results at the microeconomic as well as the macroecenomic levels are far from attaining sustainability.

Economists provide four explanations for the shallowness of environmental consciousness: (1) the availability and low cost of goods; (2) the single-household orientation in which diseconomies are associated with relatively small quantities of waste output; (3) the difficulty of choosing the abstract (environmental quality) over the concrete (consumer goods desired); and (4) the possibilities of benefiting privately from the natural environment without paying private costs.

Both the use and the destruction of the natural environment are to a considerable degree the result of legitimate behavior in keeping with existing norms and accepted values. Thus, the amount of household refuse depends on the fact that the law permits one-way packaging. The degree of air pollution depends on the right to have one's own car. On a deeper level, environmental problems result from the cumulative exercise of individual freedoms guaranteed as institutionalized rights. Moreover, environmental problems are a consequence of the fundamentally positive value our culture attaches to economic and technological innovations. Further, environmental problems are fostered by a worldview that has neutralized nature morally and reduced it to the level of a resource for man (van den Daele 1992, 3).

This catalog of reasons for the absence of a deep environmental consciousness may shed some light on possible ways to change consumer behavior to attain sustainable consumption patterns.

A variety of concepts distinguish what is good and harmless from what is bad and harmful to the natural environment. These conceptions vary according to ethical, political, and scientific understandings of the relationship between human beings and nature. According to Fischer-Kowalski, Haberl & Payer (1993), this variety of conceptions can be ordered into the following four basic paradigms:

- natural balance paradigm
- conviviality paradigm
- entropy paradigm
- poison paradigm

As Fischer-Kowalski et al. elaborate, each of the four paradigms is guided by a reference concept or orientation, and each is able to catch aspects of the possible meaning of the damage that society causes to the natural environment. The paradigms are not mutually exclusive; a specific aspect of environmental damage may occur in more than one of them. But they are in no way dependent on one another. Moreover, each has its own structure of reasoning, scientific and political traditions, and its own audience. Taken together, the four paradigms permit a scanning of what people mean when they talk about the socioeconomic system causing environmental damage. (See Figure 1.)

The four paradigms can be ordered according to the focus on the natural environment on a topographic scale with two dimensions: one dimension for the system focus (technical and physical systems versus living systems) and the other dimension for the perspective orientation (general perspective versus specific perspective). For example, the natural balance paradigm is focused on a general perspective and on physical or technical systems. Hence, thinking in terms of balancing implies a somewhat technical viewpoint. The conviviality paradigm is also focused on a general perspective, but on living systems. The entropy paradigm is focused on a specific perspective, but on physical or technical systems. The poison paradigm is focused on living systems, but with a specific perspective.

Figure 1

**Four Paradigms of Sustainability and Types
of Sustainable Consumption Styles**

focused on
a general
perspective

natural balance paradigm   **conviviality paradigm**
consumption style: *consumption style:*
voluntary simplicity   respect for life on earth
e.g., low consumption level   e.g., preservation of the
                                      countryside

focused on   focused on
physical systems   living systems

entropy paradigm   **poison paradigm**
consumption style: *consumption style:*
new technology   natural food
e.g., low-energy house   e.g., vegetarian food

focused on
a specific
perspective

Adapted from Fischer-Kowalski, Haberl & Payer 1993

It is doubtful that the four paradigms can be understood as subparadigms of the sustainability paradigm, but they obviously represent close operational implications of the sustainability concept and provide an orientation for radical environmentally conscious consumer behavior. Thus, according to the paradigms, four approaches to sustainable consumption styles can be distinguished:

- voluntary simplicity
- respect for life on earth

- natural food
- new technology

The "voluntary simplicity" style is a holistic approach to household production and consumption, with the aim of realizing a low-as-possible consumption level. The success of this style depends on achieving a sustainable household metabolism. The "respect for life on earth" style, also holistic, is focused on the living conditions of other species, with the aim of preserving countryside and habitat. This style may have a great impact on the performance of household processes as a whole. The "natural food" style is also focused on living systems but with a specific perspective—the organism of the person her- or himself. The aim of this style is to ensure the consumption of healthy food, which implies a demand for organic or vegetarian food and the promotion of organic agricultural production, and thus may lead to environmentally conscious consumer behavior in other areas of household production and consumption. Finally, the "new technology" style is also focused on a specific perspective, but on physical or technical systems. The aim of this style is to avoid reductions in standard of living through cleaner, more efficient technologies, such as energy-efficient houses and low-emission automobiles.

"Eco-pioneers" in Germany

A closer picture of these approaches to a more sustainable lifestyle can be drawn from empirical data. Various investigations suggest that the share of West German consumers adopting voluntary simplicity as a style may be as high as 10 percent of the adult population (about 6 million people). The number of "ecological radicals" in the population seems to vary according to research design—especially in terms of the method used to measure such behavior. For example, in 1993, we did an empirical case study of a group of people we called "eco-pioneers" (Doll 1993). We sought respondents from environmental association membership lists, natural-food shop customers, and the "snowball" system. The research design was based on a restrictive catalog of criteria that gave norms to various areas of household production and consumption, such as (1) activities to establish neighborhood environmental initiatives, (2) avoidance of car use whenever possible, and (3) reductions in standards for personal hygiene. We reached a total of 20 persons and selected 10 of them for data collection through a personal questionnaire of mostly open-ended questions. Though the results cannot be generalized, they do offer some interesting insights into the problems. Briefly stated, our findings are as follows:

• Most of the respondents were between the ages of 28 and 47. Half were married and living with a spouse and children. Their educational level was very high; 8 of the 10 had attended college or held B.S./M.S. degrees. Their occupational status and income followed sample population distribution patterns.

• The attitudes of the respondents were far more optimistic than pessimistic, as expressed in their feelings of capability in managing their own lives, as well as having opportunities to influence friends and neighbors. Thus, a somewhat high level of control over their own fate and of overall satisfaction with life was reported.

• Among the motivations for practicing voluntary simplicity as a lifestyle were the following: biographical events, such as pregnancy and parenthood; ethical and political beliefs; knowledge about ecological facts; a good example set by friends or family members; and a general feeling of alienation.

• Political activism for environmental initiatives and associations was reported by nearly half the respondents, but all perceived themselves as nonconformists. This general attitude may explain the political abstinence of the others.

• Most of the respondents reported that they avoid meat, prefer vegetarian food, and avoid convenience products. They pointed out that the nutritional behavior they practiced was an effort to support organic agriculture, which they understood as environmentally responsible food production.

• The respondents reported (and informal observation revealed) a well-below-average possession and use of household appliances. Half the respondents had neither a TV nor a VCR. They used refrigerators and washing machines but did not use dishwashers or small kitchen appliances, such as electric knives and electric can openers. Some of the respondents who were homeowners had appliances for collecting rainwater and using solar energy.

• Transportation behavior was reported as a somewhat critical issue. Only 3 of the respondents had no car.

In summary, "voluntary simplicity" can be understood as a holistic approach to a sustainable lifestyle and consumer behavior. The main trait of this style can be seen in the high priority given to *not* consuming certain goods at all and, thus, contributing to the conservation of natural resources.

In contrast, consumers who practiced the "natural food" consumption style gave high priority to the avoidance of meat and nonorganic agricultural products. This style covers various forms, and the reasons are also varied for preferring vegetarian food. The main motivations are that organically grown vegetarian food is healthier and better for the environment (Brombacher & Hamm 1990; Keler, Kutsch & Köpke 1994; Schilling 1994). According to an estimate by the Deutsche Gesellschaft für Ernährung (1992, 68), this segment of consumers may account for from 5 to 7 percent of the West German population.

The theoretically based assumption that the natural food consumption style implies more far-reaching environmentally conscious consumer behavior (Piorkowsky 1988, 18-24) has been demonstrated by a household survey conducted in 1989 (Huber, Piorkowsky, et al. 1989). The data were collected in 109 households in Hamburg, one of the largest cities in West Germany, via a personal questionnaire. Of the respondents, 55 percent practiced a somewhat "natural food" consumption style, especially the vegetarian style; 45 percent preferred conventional food.

Our findings, in short, are as follows:

• Of the respondents practicing a "natural food" consumption style, nearly 80 percent were female, and nearly 50 percent were between the ages of 25 and 39. Household size and occupational status were distributed as in the sample population.

• A statistically significant impact of the "natural food" consumption style on environmentally conscious consumer behavior has been evaluated in the following areas: (1) less use of softening agents to wash clothes; (2) more special appliances and usages to save water, especially when taking a shower instead of a bath; (3) more avoidance of packaging materials for consumer goods; (4) more opportunities to separate household waste by components; and (5) more bicycle riding and walking in everyday life, as well as on vacation.

Although our sample was not representative of the West German population with its observed subgroups, the results of the household survey support the assumption that there is a positive correlation between "natural food" consumption and more far-reaching environmentally conscious consumer behavior. Thus, preferring natural food implies, or leads to, a more sustainable lifestyle.

The same conclusion can be applied to a segment of environmentalists and nature conservationists. Since the mid-1980s, some 1,500 environmental initiatives or associations with more than 5 million members have come into existence (Simonis 1991, 5-6). Empirical evidence suggests that a considerable portion—if not all—of the members of such initiatives and associations are in many ways identical with consumers who practice the "respect for life on earth" style.

Supporting data are available from a 1991 case study by Christmann of environmentalists and nature conservationists in a medium-sized town in southern West Germany (assumed to be Constance). Data were collected by informal observation of 62 environmental initiatives and environmental associations in this town and by interviews with 40 activists. The findings are as follows (Christmann 1992, 202-8):

• Nearly half the respondents were students. Most of the others were employees in areas of teaching, research, and social work. The educational level was extremely high; some respondents had two B.S. degrees. Only a few of them were housekeepers or retired.
• The main purpose and motivation of the activists was to instill knowledge and provide public information about the environment.
• A positive impact on the performance of household production and consumption was evaluated with respect to transportation and leisure activities, especially avoidance of travel by car and airplane and refraining from skiing.

Some note must also be made of the segment of consumers who practice the "new technology" consumption style. Scarce data are available to validate the assumption of more far-reaching environmentally conscious consumer behavior within this group and within subgroups, respectively. Of the subgroups, users of low-energy houses are a major subject of energy policies, as well as of related energy and social research (Sandtner 1993). The Frauenhofer Institute has pointed out that using solar energy requires additional energy from other sources or behavioral changes to compensate for the variable solar supply. Thus, users get a feeling for the dynamics of the natural cycle, which may lead to a more sustainable household-management style (Prose, Hübner & Kupfer 1993, 19).

A 1993 case study by Kupfer on users of photovoltaic units for collecting solar energy revealed the following:

• All of the respondents were homeowners. Half were between the ages of 40 and 53; on average they were 48, and 60 percent lived with children. The educational level was high: 60 percent held B.S./.M.S. degrees. The occupational status and income were also high. Most of the respondents were self-employed or employees in the areas of teaching, engineering, business administration, and health service.

• Among the respondents' motivations for using photovoltaic units were the following: (1) environmental consciousness—doing something beneficial for the environment (90 percent), (2) economic purpose—to spend less on energy (52 percent), and (3) inquisitiveness about new technology (27 percent).

In conclusion, the empirical data from Germany have nourished the hope for a new model of sustainable consumption. Thus, research on consumer behavior must be continued.

References

Ayres, R.U. 1994. "Industrial Metabolism: Theory and Policy." In R.U. Ayres, E.U. Simonis, eds. *Industrial Metabolism. Restructuring for Sustainable Development.* Tokyo, New York, Paris, 3-20.

Boulding, K. E. 1968. "The Economics of the Coming Spaceship Earth." In Jarret (below), 3-14.

Brombacher, J. & U. Hamm. 1990. "Für alternative Butter, Sahne und Bananen müssen Bio-Haushalte kräftig zahlen." *Rationelle Hauswirtschaft 27:* 10,10-12.

Christmann, G.B. 1992. "Wissenschaftlichkeit und Religion: Über die Janusköpfigkeit der Sinnwelt von Umwelt und Naturschutzern." *Zeitschrift für Soziologie 21,* 200-11.

Deutsche Gesellschaft für Ernährung. 1992. "Ernährungsbericht 1992." Frankfurt am Main.

Doll, E. 1993. "Ökopioniere. Eine empirische Pilot-Studie." Unveröffentlichte Diplomarbeit an der Landwirtschaftlichen Fakultät der Rheinischen Friedrich-Wilhelms Universitat: Bonn.

Fischer-Kowalski, M., H. Haberl & H. Payer. 1993. *Economic-Ecological Information Systems. A Proposal.* Science Center for Social Science Research, FS II 93-406 Berlin.

Frauenhofer-Institut für Solare Energiesysteme: "Sonnenstrom von tausend Dächern." Freiburg.

Georgescu-Roegen, N. 1970. "The Entropy Law and the Economic Problem." In: N. Georgescu-Roegen *Energy and Economic Myth*. Institutional and Analytical Economic Essays. New York: 53-60.

Huber, L. & M.B. Piorkowsky et al. 1989. *Empirische Studie zum Umwelt und Ernährungsverhalten*. Fachübergreifendes Projekt der Arbeitsgruppe Ernährungsverhalten des Fachbereichs Ernährung und Hauswirtschaft der Fachhochschule Hamburg im Sommersemester 1989. Unveröffentlichter Untersuchungsbericht.

Jarret, H. ed. 1968. *Environmental Quality in a Growing Economy*. Baltimore, MD.

Keler, T., T. Kutsch & K. Köpke. 1994. "Produkte des kontrolliertbiologischen Anbaus im konventionellen Lebensmitteleinzel-handelunter besonderer Berücksichtigung des Konsumentenver-haltens." *Hauswirtschaft und Wissenschaft 42*, 51-58.

Kupfer, D. 1994/95. Analyse der Nutzer solarthermischer und photovoltaischer Anlagen. Report. Unternehmensberatung Kupfer. Kiel.

Piorkowsky, M.B. 1988: "Umweltbewusstsein und Verbraucher-verhalten." In: M.B. Piorkowsky & D. Rohwer *Umweltverhalten und Ernährungsverhalten*. Hamburg, 7-87.

Prose, F., G. Hübner & D. Kupfer. 1993. "Nordlicht. Zur Wirkung einer Klimaschutzkampagne." Draft. Projekt Energiesparen. Institut für Psychologie der Christian-Albrechts-Universität: Kiel.

Sandtner, W. 1993. "Zum Bund-Länder-1000-Dächer-Photovoltaik-Programm." In: *Elektrizitätswirtschaf 92*, 1537-1540.

Schilling, D. 1994. "Verbraucherbefragung zu den Erzeugnissen des ökologischen Landbaus in Ostdeutschland." *Hauswirtschaft und Wissenschaft 42*, 59-64.

Simonis, U.E. 1991. *Environmental Policy in the Federal Republic of Germany*. Science Center for Social Science Research, FS II 91-403 Berlin.

van den Daele, W. 1992. *Research Program of the Section Norm-Building and Environment*. Science Center for Social Science Research, FS II 92-301 Berlin .

World Resources Institute. 1992. *World Resources 1992-93. Toward Sustainable Development*. New York: Oxford.

Chapter 5

Linking Teacher Education and Environmental Education: A European Perspective

C. R. Oulton and W. A. H. Scott
Environmental Education Research Group
University of Bath, School of Education
United Kingdom

This chapter reports on some of the outcomes of the first year of the Environmental Education into Initial Teacher Education in Europe (EEITE) project. Over the last three decades it has been widely accepted internationally that EE[1] should be a component of the formal educational system (IUCN 1970; UNCED 1992). Alongside this assertion often comes a recognition of the need for preservice and inservice teacher education programs that allow individual teachers to contribute to EE in schools (IUCN 1971; UNESCO-UNEP 1976; UNESCO 1978; UNESCO-UNEP 1987; UNESCO-UNEP 1990).

The European Union supports this view. In "Towards Sustainability: A European Community Program of Policy and Action in Relation to the Environment and Sustainable Development" (EU 1993), education is once again recognized as a key element:

> The importance of education in the development of environmental awareness cannot be overstated and should be an integral element in school curricula from primary level onwards. (106)

The document calls for a speeding up of the implementation of a resolution made earlier by the Ministers of Education of member states which resolved that "all sectors of education should make a contribution to carrying out environmental education" (EC 1988).

Despite these repeated assertions, in many countries only limited progress has been made at school and teacher education levels.

Although progress has been made in the training of personnel for EE over the past decade, and a number of national initiatives have shown promise, considerable needs remain unmet in all regions of the world. There is a shortage of teachers for EE at all levels. A number of strategies have been devised to provide environment-linked teacher training, and their effectiveness has proved highly variable (UNESCO-UNEP 1988).

A number of attempts have been made to define the competencies of environmentally educated teachers (UNESCO-UNEP 1990). Elsewhere (Oulton & Scott 1995), we have argued that long lists of competencies may be useful in defining what is to be achieved, but they say little about how to achieve it.

It is against this background that the European Union's EEITE[2] project was initiated. It was developed as one aspect of the work of the Association for Teacher Education in Europe (ATEE) Working Group 17: Environmental Education and Teacher Education.

ATEE is the leading organization promoting collaboration, research, and development for teacher education in Europe. Within the organization are 19 working groups, each focusing on a particular aspect of teacher education. Members of the working groups meet together at the annual ATEE conference. Each conference has a theme that is addressed through the main conference program. In addition, a series of working group sessions are also scheduled. Working Group 17 was formed in 1992. Its purposes are the following:

- to establish a regular forum for teacher educators with an active interest in EE
- to promote research on EE in the context of teacher education
- to promote publications in the field of EE
- to develop and strengthen links with existing networks, especially EU and UNEP projects
- to relate European EE issues to a global context
- to stimulate the development and implementation for curricula of both preservice and inservice teacher education

The EEITE Project

The EEITE project was developed to meet most of these aims. The project's prime objective is the coordinated development of teaching units to be used in teacher education aimed at the preparation of teachers for EE in primary, secondary, and vocational education.

Specifically, the project has set out to do the following:

- create an inventory of current practice in preservice teacher education programs (primary, secondary, and vocational) across member states
- review implementation strategies in EE in each country and their relation to preservice courses
- develop a strategy to implement EE into preservice courses in each EU country, taking account of phases, curriculum constraints, course organization, and current practice
- establish criteria whereby implementation strategies can include issues relating to conceps, values, and taking action

The project began with Country Project Representatives (CPRs) from 11 EU countries meeting together with a group of EE experts from across Europe in the first "Summer University," held in Brussels in October 1993. At this meeting, each CPR presented a commentary on the state of EE in her/his own country, together with an evaluation of EE in preservice teacher education programs. An analysis of these CPR reports can be found in Brinkman & Scott (1994). The key issues raised in that publication give an indication of the complexity of the situation facing a group trying to establish common activities within teacher education programs across a group of countries.

Across the individual institutions and countries represented in the project, a range of factors may be identified that influence the provision for EE in preservice teacher education programs. These factors include the following:

Preservice Teacher Education

- the nature of preservice (if any)
- the number of alternative pathways to qualified teachers status (QTS)
- national frameworks and arrangements for course scrutiny and approval
- the influence of school curricula on preservice teacher education courses
- the freedom that tutors have in delivering curricula

Environmental Education

- its significance within any agreed (national) school curriculum
- any consensus about its national importance
- the interest and awareness arrangements for course managers and tutors, and their skills and competencies in this area

- links between NGOs and preservice teacher education courses
- national policies and imperatives

Brinkman & Scott (1994) suggest that the diversity generated by this multiplicity of factors may be conceptualized as being multilayered. A selection of examples from his analysis will illustrate the point.

Layer 1: National Circumstances

In some European countries (for example, Greece and Italy) no preservice teacher education courses per se are provided for prospective secondary school teachers. Where courses are provided, they differ in length from 26 weeks to five years. Within any one country, a variety of types of teacher education systems may be found. In England and Wales, there are currently nine routes to QTS.

Layer 2: Institutions

In most of the project countries, teacher education institutions work within some externally provided framework. In Greece, for example, it is a national framework (see Gravanis, this volume). In Germany, it is regional (see Böhn, this volume). Other countries, such as the Netherlands and Denmark, have decentralized systems. The extent to which EE is mentioned in these frameworks also varies. In Germany, EE is seen as a fundamental prerequisite of the required competency of a teacher, whereas in Denmark, there is no mention of it in the state's objectives for the preservice training for teachers. A further variable here is the "degrees of freedom" that institutions have in interpreting national frameworks and the extent to which aspects of courses such as EE are scrutinized externally. In England and Wales, for example, this is done by Her Majesty's Inspectors.

Layer 3: Courses, Management, and Course Teams

A range of options exists for incorporating EE into courses. These include permeation of EE into, for example, subject didactic programs; short introductory courses for all students; option courses for small numbers of students and mainstream courses for EE specialists. The mode adopted by a particular course team will often depend on the ability of interested individuals to promote their area of interest in competition for curriculum space and other resources.

Layer 4: Tutors

Because tutors normally have some degree of freedom to interpret and modify course elements, seemingly common components of a course may be presented very differently from one tutor to another. Tilbury (1993) suggests targeting tutors in any attempts to promote EE within preservice teacher education programs.

Layer 5: Schools

There is diversity in the extent to which schools play a part in initial teacher education. For example, in England and Wales, France, and the Netherlands, school experience is an integral part of the course, whereas in Portugal, school experience follows the preservice training program. In some systems, school experience is minimal. Where schools have a role to play, the degree to which they are able to support work on EE varies depending on, for example, the emphasis given to it in the school curriculum and the interpretation of EE made by supporting teachers.

Layer 6: Individual Students

In most European countries, it is accepted that all teachers have at least some responsibility for delivering EE within the curriculum; in some countries, teachers of particular subjects (such as science and geography) are recognized as playing a leading role. Students arrive for preservice training, with or without their own experience or EE in their own schooling and in their degree studies. As yet, insufficient attention has been given to the need to differentiate EE training programs for prospective teachers of, say, secondary mathematics from those of science, or secondary school from primary school teachers.

Meeting the Challenge of Diversity

Faced with this range of diversity across the European Union, any project working at the European level needs to avoid prescription in either the content of a program of study or the implementation strategy employed in a particular institution. It is clearly inappropriate to propose any pan-European module for EE in preservice teacher education programs.

While avoiding being prescriptive, the members of EEITE agreed on the following organizing principles for the project:

- As a result of preservice teacher education programs, novice teachers should be both willing and able to make a contribution to EE through their own work with learners.
- Being willing suggests that the teachers understand the importance of EE and have a personal commitment to it that is both practical and intellectual.
- Being able suggests that they have a repertoire of strategies for managing change and an innovative curriculum on which they can draw in cooperation with others.

The EEITE project recognizes that these are ambitious aims, and in order to achieve them, preservice programs will need to contain two elements. For the sake of clarity, these elements are listed here separately. This should not be taken to mean that these will necessarily be separated in practice. Tutors will have the responsibility for deciding the interrelationship between these (and other) elements for themselves and for determining the patterns of organization and support that their development work will have. Rather than stifle innovation, it will be necessary to encourage diversity and monitor practice to gain insights into the transferability of particular approaches and program design between institutions. The two elements are these:

(1) Aims and Practice

- consideration of the aims and practice of EE, particularly as it relates to compulsory schooling
- examination of curriculum practice and extracurricular opportunities and the desired learning outcomes associated with these
- identification of those characteristics that mark out curriculum as contributing actively to EE
- exploration of particular strategies and approaches that can be employed in EE

(2) Personal Experience in Environmental Education

- working with teachers and children in schools on small-scale activities
- evaluating this practice and building on the foundations laid through reflection and systematic planning
- in particular, evaluating the effects of this practice on both their own and children's awareness of the possibilities and priorities of EE

It is necessary to emphasize the incremental and iterative nature of such developments and the consequent necessity of taking a small-step approach, coupled with a focus on the management of intervention and change.

In addition, the EEITE project evolved a number of characteristics that each CPR, in the academic year 1993-94, would use to develop an EE component within the preservice teacher education program in his/her own institution. The characteristics are the following:

- in part at least, a local focus drawing from, and contributing to, expertise and awareness in the local community
- integration into preservice teacher education programs rather than being an addition
- a clear set of aims and desired learning outcomes related to the goals of the preservice program
- being action-oriented, with novice teachers who are involved in the planning, implementation, and evaluation of the work and will be encouraged to have an individual commitment to reflection so as to build the experience into their own professional development
- the use of values and attitude development as key features
- the ability to share the processes and outcomes of the work with other subject groups
- an interdisciplinary approach, involving more than one subject area or curriculum focus
- a dual focus, with tutors and teachers working with novice teachers who, for their part, work with students in schools

The outcomes of these individual development projects were reported by CPRs at the second project conference in Spain in July 1994. An interim report was made to the European Commission in the autumn of 1994.

The UK CPR Project at Bath

This section provides an overview of the way in which the principles of the EEITE project outlined above were incorporated into one institution's preservice course. This is followed by a discussion of some of the issues raised by this innovation.

The program was run for 195 novice teachers as part of a 36-week Postgraduate Certificate in Education (PGCE) course for secondary school teachers in the subject areas of English, history, mathematics, modern foreign languages, science, and physical education.

The project comprised three 2.75-hour workshops held on consecutive weeks during January and February 1994. The workshops, under the title "Personal and Social Education" (PSE), dealt with the cross-curricular themes in the National Curriculum for England; these are careers, health education, citizenship, and economic and industrial understanding, as well as EE. The workshops were a compulsory element of the PGCE course. Within the workshops, as explained below, participants were given an element of choice over which theme they "specialized" in. The PSE workshops ran in tandem with a second set of workshops relating to "Equalizing Opportunities." These workshops focused on issues of gender, race, sexuality, and pupils with special educational needs. Half of the course did one of the sets of workshops for the first three weeks and the other set in the subsequent three weeks.

The overall purpose of the PSE workshops was to raise the awareness of novice teachers of the objective of the themes and, through an investigation of the way in which themes are handled in their home schools,[3] to offer them some strategies for incorporating the themes into their work as teachers. As a CPR for EEITE, Oulton attempted, through his work as coordinator of the workshops, to attempt to ensure that, although EE was only one of the five themes, key aspects of EE would permeate the entire program.

The workshops were planned by the five tutors who delivered them. Each had particular knowledge of one of the themes. As coordinator of the workshops, Oulton suggested to the planning group that some of these objectives could be met by involving the novice teachers in a project to explore the local issue of traffic problems in Bath. Oulton argued that traffic offered a context for work in each of the five themes, provided an interesting and topical issue to consider, and demonstrated the active-learning, locally focused approach that good education in each of the themes involved. Some colleagues were not convinced that traffic provided the most fruitful context and suggested they would personally be more able to support work on other issues. It was therefore decided that each tutor would choose a social issue for his/her group to work on. The issues chosen were traffic in Bath, AIDS, and teaching about development issues (two tutors). The fifth tutor allowed novice teachers to identify an issue for themselves. The detailed format and objectives for the workshops were agreed upon.

Oulton was responsible for writing materials for the first session that introduced the themes and some of the underpinning ideas that the project team felt were important. This first session emphasized the following:

- the underlying elements of the themes, e.g., making decisions about myself; making decisions about society; views of the the importance of values; and thinking about desirable futures
- common pedagogy of the themes, e.g., problem solving; decision making; values analysis and clarification; data handling; role play; and using real-life contexts for learning
- interrelatedness of the themes, e.g., the need to consider aspects of industrial and economic understanding or health education when considering an environmental issue

At the beginning of the second session, novice teachers spent some time with specialist tutors exploring one of the themes. They then returned to their first tutor, who introduced the topics on which their group had chosen to focus. In Oulton's case, the traffic problems in Bath, the session was run as a role play in which the novice teachers took part as the members of an imaginary pressure group. This session was planned with the help of an environmental education officer who is employed by the local Council Architects' Department. As well as providing very helpful advice on developing the role play, she had access to a wide range of recent data. She also joined in one of the role-play sessions. The briefing sheet for novice teachers was as follows:

The Social Issue

You are a member of BRACT (Bath Residents Against Congested Traffic), a group formed at a public meeting held to discuss the ever-increasing local traffic problems. Traffic issues raised in that discussion included limited parking for shoppers, road safety, the time taken to get into the center of Bath from the London Road end, the lack of facilities for cyclists, complaints about the parking scheme, concern about air pollution, the need for a bypass, a proposal to make Broad Street a traffic-free shopping district, a request for a pedestrian crossing on London Road, and the need for park-and-ride facilities. The wide range of issues raised resulted in a rather diverse and unstructured debate about what BRACT's priorities should be. It was generally agreed that more information was required. A number of issues were selected, and working parties were set up to investigate the topic and report back at the next meeting. Each member of one of the working parties would be given a separate working-party briefing sheet.

In summary, the health education group was given data on health problems caused by air pollution and local pollution levels and was asked to find out what people in Bath knew about the problem. The careers group looked into the possible impact on jobs of the pedestrianization of part of the city. The citizenship group interviewed a traffic engineer from the Council to find out how citizens may influence the road-planning process. The economic and industrial understanding group looked at the impact on shops of closing a main shopping street to traffic. The environment group monitored noise pollution and looked at the visual impact of traffic on city esthetics. Completing these activities took most of one session.

The first part of the final morning of the program was run as a role play of a BRACT debate. Each working party had to make a presentation of its findings, and the group had to discuss what the next phase of the campaign should be. Finally, the role play was debriefed, and the whole project discussed. During the BRACT meeting and the subsequent discussion, the group was joined by a member of the local council who had responsibility for traffic in Bath and who was thus not only able to join in the role play debate as himself but also offered comments during the debriefing session.

Issues Arising from the Evaluation of the Program

Data from a number of sources were used to evaluate the project, including the responses of tutors, novice teachers' comments in discussion at the end of the workshops, a questionnaire survey, and a second survey after the novice teachers had completed a block of teaching. This section discusses some of the issues raised in the evaluation.

Interpreting the data was difficult since EE was, by design, integrated with other themes in the program. On top of this, the timing of the workshop, a factor over which Oulton had no control, created problems for novice teachers—some of which were reported in the questionnaires and some of which resulted in some novice teachers absenting themselves from the workshops. These whole course issues included: First, the fact that the workshops started at 9:15 although university sessions normally start at 10:15 to accommodate novices who have trips of over one hour to get to the university. Second, the workshops were run at a time when most novice teachers were also meeting a number of coursework deadlines—thus, the "extra" course element was not universally appreciated. Third, the approach adopted by tutors running the "Equalizing Opportunities" workshops was criticized by some novice teachers who attended them in the first three weeks, so part of the cohort attending the second set of workshops in the following three weeks arrived with a rather jaundiced attitude.

Although these issues may be considered somewhat trivial, they illustrate the importance of avoiding any lowering of the status of a course element such as this and the need to avoid its being perceived as a "bolt-on" extra. Work on EE needs to be embedded in the course program, even if it is delivered through special elements such as the workshops. Ways need to be found to forge links between such program and mainstream elements of the courses. Alternatively, elements of the workshops need to be transferred to mainstream course elements. For example, tutors at Bath have recently agreed that a workshop on teaching about controversial issues should be incorporated as part of the science didactics program.

Given the limited time available for the workshops, our main aims were to inform novice teachers about the five themes in a general sense and to give them some knowledge about one theme. To a large extent, the responses suggest that this was achieved. Many novice teachers stated that they were more aware of the themes and that they now had some ideas for how they might incorporate the themes into their teaching.

The idea that focusing on principles across the themes and the common ground among the themes would avoid the need to cover all of the themes at a superficial level does not seem to have been understood by the novice teachers. The fact that many novice teachers wanted to learn more about themes other than the one they specialized in can be considered as a positive outcome, even though it was expressed as a negative point in the evaluation.

In a questionnaire completed after a block of teaching experience, half of the novice teachers responding had managed to include some work on at least one of the themes in their teaching. However, in looking across the different subject groups, some marked differences were noted. For example, 73 percent of novice teachers of science had done some work on the themes compared to only 24 percent of novice teachers modern foreign languages (MFL). In fact, fewer than 1 percent of novice teachers of MFL managed to include work on any of the themes in their subject teaching. The pattern across all subject groups needs to be looked at carefully and should raise questions about the way in which the themes are presented to the novice teachers. There seems to be little point in trying to convince them that their subject area has a contribution to make if this is not supported by their experience in the school. It may be that some subject groups need to focus on particular themes. The survey certainly reveals a link between, for example, physical education and health education, with 88 percent of novice teachers of PE reporting work in this area.

These course development issues raise questions that need to be addressed through research activities. For example, if it is assumed that novices need to develop their competence in EE in the classroom, to what extent are teachers able to support them in this work? If teachers themselves need support, precisely what support is needed, and how may that support be effectively given? What role might university tutors play in supporting novices and tutors? Can this be achieved through collaborative work involving tutors of novices, teachers, and pupils working together on an environmental issue?

The Second Phase of EEITE

EEITE is now in its second year, and CPRs are engaged in developmental work at two levels. First, each CPR is charged with extending the work within his/her own institution. This may include working with more novice teachers, working with other colleagues, or working at greater depth. Second, bilateral and trilateral research programs between CPRs have been established. Each collaborative program focuses on a mutually agreed upon set of research questions arising from the work in the first year. The outcomes of these research activities will be published at a later date.

Notes

1. "Environmental education" was the term used when the project reported in this chapter was initiated. More recently, the term "education for sustainability" has been used. While we recognize that the two terms are not synonymous, most of the arguments in this paper could be applied to the education of teachers for both goals.
2. EEITEE is funded by the Directorate General XI of the European Commission.
3. Novice teachers spend two thirds of the PGCE course working in schools. At the time of the workshops, novices were spending two days a week working in their home school.

References

Brinkman, F.G. & W.A.H. Scott, eds. 1994. "Environmental Education into Initial Teachers Education in Europe (EEITE): the State of the Art." ATEE Cahier No. 8, Brussels, Association of Teacher Education in Europe.

European Commission. 1988. "Resolution of the Council and the Ministers of Education meeting with the Council, Environmental Education." (88/C177/03), EC, 24 May.

European Union. 1993. *Towards Sustainability: A European Community Program of Policy and Action in Relation to the Environment and Sustainable Development.* Brussels: Commission of EU.

IUCN. 1970. *International Working Meeting on Environmental Education in the School Curriculum.* Paris, UNESCO.

IUCN. 1971. *International Working Meeting on Environmental Education in the School Curriculum.* USA, IUCN.

Oulton, C.R. & W.A.H. Scott, 1995. "The Environmentally Educated Teacher: An Exploration of the Implication of UNESCO-UNEP's Ideas for Preservice Teacher Education Programmes." *Environmental Education Research, 1* 2:2.

Scott, W.A.H. 1994. *Diversity and Opportunity Reflections on Environmental Education within Preservice Teacher Education Programmes Across the European Union.* In Brinkman & Scott (above).

Tilbury, D. 1993. *Environmental Education: Developing a Model for Initial Teacher Education.* Unpub. diss. University of Cambridge.

UNCED. 1992. United Nations Conference on Environment and Development, Promoting Education, Public Awareness and Training, Chapter 36, *Agenda 21*, Conches: UNCED.

UNESCO-UNEP. 1976. The Belgrade Charter, *Connect, 1* :1, 1-8.

UNESCO 1978. *Final Report: Intergovernmental Conference on Environmental Education.* Paris: UNESCO.

UNESCO-UNEP. 1987. *Strategies for the Training of Teachers in Environmental Education.* Paris: UNESCO.

UNESCO-UNEP. 1988. *International Strategy for Action in 1990.* Nairobi/Paris: UNESCO-UNEP.

UNESCO-UNEP. 1990. Environmentally Educated Teachers: The Priority of Priorities? *Connect 15* :1, 1-3.

Chapter 6

Happiest but Least Cheerful: The Paradox of Population and Education Policies in China

Gaoyin Qian and Tiechi Huang
Lehman College
The City University of New York

The notion of "keeping harmony with nature" in Chinese philosophy is ubiquitous in Chinese culture, from paintings, poetry, and architectural designs to their intention to "solve the various problems of human adaptation to material reality" (Leys 1996, 30). To the Chinese mind, the environment is part of "material reality," but it at the same time contributes to spiritual reality. T'ao Ch'ien (also known as Tao, Yuan Min, A.D. 365-427) was one of the literary giants in Chinese history, who took pleasure in his garden and contemplated the beauty of nature. He wrote in his famous piece of poetic prose *Homeward Bound I Go:*

> Wealth I want not; honor I desire not; the Heavenly Palace is beyond my reach! I would stroll alone on a bright morning, or plant my cane and till the ground.

> Or I would mount the eastern hill and sing my song, or visit the clear brook and make my poems. Thus would be content to live and die, and be glad indeed to accept the Will of Heaven. (Chai & Chai 1965, 32)

This prose is known to almost every educated Chinese person. Recent developments in China have severely tested this ancient reverence for nature. As noted by Cheng:

> The modern [M]andarin translation for *environment* is *huan-chin*, meaning "world of surroundings." This translation apparently reflects the surface meaning of *environment* correctly. But when embedded in the contexts of Chinese philosophy and Chinese cultural consciousness the "world of surroundings" does not simply denote individual

things as entities in a microscopic structure; it also connotes a many-layered reality such as heaven and earth in a macroscopic enfoldment. This "world of surroundings" is generally conceived as something not static but dynamic, something not simply visible but invisible. It is in this sense of environment that we can speak of the *tao* as the true environment of man: the true environment of man is also the true environment of nature or everything else in nature. (355)

As in other modern nations, the Chinese government has had to deal with their country's "carrying capacity." This has led to resolving the tension between pollution and environment with the ongoing "one child per family" program which has been scrutinized and criticized by the West. Whether or not the program will eventually bring "harmony with nature" remains controversial. This chapter examines the influence of Chinese population policies on environmental, educational, and social issues.

A Full House of Children

For centuries, Chinese parents enjoyed living with "a full house of sons and grandsons." The Chinese love of children is so deep that even the formation of certain characters in the writing system symbolize this aspect of the culture. For example, the character for "goodness" or "something good" is coined by the two radicals "daughter" on the left and "son" on the right. However, as the population grew, land became increasingly scarce because new houses were built and new farms were created. Due to rapid population growth during the 1950s and 1960s, the Chinese government started a campaign in the late 1970s to convince each family to have no more than one child. This historic change from large to "only child" families has challenged beliefs about children that the Chinese people have held for centuries. Educators and theorists are concerned about the effects of the one-child program on the only child who grows up in an environment without siblings, uncles, aunts, or cousins. Nevertheless, government at all levels has resorted to incentives, disincentives, and social pressures on individual families to achieve the goal to stop the growth of the population at 1.2 billion people by the end of the 20th century. However, the 1.2 billion population figure has already been reached as of 1996.

In present-day China, children are regarded as the happiest people, but children in pre- and elementary schools are also the least cheerful. They are the happiest because they are the center of the family and "little emperors." Parents, grandparents, and the public do their best to provide them with resources, to let them enjoy different kinds of privileges, and to keep them away from doing manual labor. Children are also the least cheerful because they are made to study at a very early age. They must carry a backpack that weighs more than 10 pounds to school. In addition to studying in school for six hours daily, they have to preview for the next day's lessons, review what they have learned, and do extra homework assigned by both teachers and parents. In order to finish these assignments, they usually work until midnight. Even on Sundays they are required by their parents to spend several hours in special-interest programs related to music, art, or science.

Why are the children—who are regarded as China's happiest—the least cheerful? How have China's population policies contributed to this paradoxical phenomenon among Chinese children? There has been an increased interest in examining the influence of cultural, historical, and environmental factors on a child's mental development and learning (Heath 1982; Wertsch 1993). In the following sections, culturally derived norms, educational policies, and societal expectations will be examined in order to offer insights into the issue of EE in a country that is already adopting population control policies.

Culturally Derived Norms

Traditionally, the Chinese respect the culturally derived norms that "nothing is superior to study" and "those who do well in studies become officials." These culturally derived norms that originated in Confucianism have molded the Chinese people's beliefs about and their attitudes toward schooling. Chinese parents expect their children to perform well in school because success in learning will eventually lead to high social status and wealth. Historically, people who held official government positions enjoyed privileges (e.g., power and wealth). Most of the official positions in government offices went to those who excelled on the Imperial Examination administered at different levels. The Imperial Examination system was established in A.D. 587 in the Sui dynasty. Although the examination was abolished in 1905 by Emperor Guangzu, its influence on learning and schooling is still evident in the Chinese educational system (Li 1996).

Culturally derived norms also interpret what is meant by going to school and receiving an education. No matter who governs and which political group stays in power, being politically correct, inheriting cultural traditions, and following social norms have been the expectations of students in school.

Despite social changes after the 1949 Revolution, the essence of education is almost the same as the educational philosophy advocated by Chu Hsi, who lived in the Sung dynasty more than 800 years ago. The Confucian philosopher emphasized that the purpose of letting people receive education is to have "a balanced style of living" or "to live in harmony with the moral law" (Cotterell & Morgan 1975). Chu Hsi believed that evil was the result of neglect and the absence of proper education. His idea of re-educating people who have been led astray by the distortions of the society in which they live was also reflected in the policy during the Cultural Revolution that those "who have taken the capitalist road" should be re-educated.

The emphasis on being "politically correct" in education appeared to contradict the notion of "keeping harmony with nature" in Chinese philosophy. On the one hand, Chinese artists, writers, and architects impressed the world with their accomplishments in intertwining their moods, feelings, and ideas with nature through their paintings, poetry, and architectural designs (Altman & Chemers 1980). On the other hand, writers and artists were fearful that they might be politically persecuted because their artworks and subjects in writing were not in line with government policy. Instances of people who were punished or executed for their "political incorrectness" have been documented in many history books. Even students in school had to be cautious about what they wrote. For example, in the 1970s, there was a "mass denunciation" movement that was set off by a student's composition titled "All Bright?" The student made a contrast between a sad and dark corner in the real world and the "bright life" portrayed on the screen (Li 1996). As a result, literal learning has become a safe way to relieve teachers and parents from such concerns and to ensure that students will be "healthy and politically correct" in school. Literal learning refers to a way of learning in which the recitation and imitation of classics and conformity to traditions are emphasized.

The rigid control over the content and conformity to traditions inevitably led teachers to stress structures, traditions, and moralistic values in education instead of fostering students' flexible learning and critical-thinking abilities and encouraging their self-expression. Historically, the Imperial Examination required that candidates recite passages from the classics, imitate the styles and techniques of great scholars, and compose essays to extol the ruler (Huff 1993). The examination tended to make candidates overlook their creativity and self-expression. Oriented by the government-controlled Imperial Examination, the subject matter children learned in school was focused on Confucian classics, poetry, and official histories.

Literary learning was valued highly even during the Cultural Revolution from 1966 to 1976, although the notion that "nothing is superior to study" was constantly under attack and criticized. At the peak of the Cultural Revolution, people were expected to recite Mao's quotations and works verbatim. Things in China now have changed dramatically because of economic reforms since the early 1980s, but schools, parents, and teachers still overemphasize the importance of literal learning from the classics, great scholars, and government-approved textbooks and staying in line with the government so as to be politically correct (Li 1996). As a result, children in school perform better on tests that require memorization than in academic tasks that require critical thinking and creativity, such as solving environmental problems.

Educational Policies and Parents' Unfulfilled Dreams

China's current educational policies have also contributed to the paradoxical phenomenon of "being the happiest but least cheerful" in elementary education. Because of limited resources, only a small proportion of students can have an opportunity to pursue undergraduate study. The selection of the best among students as early as children start kindergarten has caused a chain reaction in elementary and secondary schools. Schools have to compete with each other to send their best students to "key elementary and secondary schools" that are funded with more government resources, equipped with better teaching facilities, and staffed with more competent faculty. Elementary school teachers are judged by the number of their students who can succeed in going to the key secondary schools, whereas secondary school teachers are evaluated on the basis of their students' success in going to college. School administrators' high expectations have led teachers to assign extra amounts of homework to students. As a result, children must do a lot of worksheets and are expected to learn more than the curriculum requires.

Parents' unfulfilled dreams have also burdened the only child with extremely high family expectations. China went through cultural and political turmoil from 1966 to 1976, during which time schools were literally closed. People who are now in their forties were sent to remote rural areas to be brainwashed ("re-educated") through manual labor. Only a select few could pursue higher education. The majority of school-age young people were deprived of formal education during that time. There has been a strong feeling of anger and a sense of being neglected among these middle-aged people who are now the parents of one child. They regret that they did not have the opportunity to pursue further learning but expect their only child to perform extremely well in schools so that their unfulfilled dreams can be realized. On the one hand, parents are very strict with the child in his/her schoolwork. They are very much involved in the child's learning by investing time and money to ensure that the child gets the best education possible. On the other hand, parents and grandparents spoil the child because they are willing to provide anything to please the child so long as she/ he performs well in school.

Chinese culture and history have developed norms that influence how schooling and learning are conceived. Children have to live up to the expectations of their parents who wish their children could realize their own unfulfilled dreams by being absorbed in literary learning in order to excel on the examination. Children who are "professional learners" are more likely to succeed in going to key schools and college but at the expense of their "joyful childhood." There have been many instances of success due to children's hard work, driven by high societal expectations and culturally derived norms. For example, preschoolers are able to recite classical poems and passages, and middle- and high-school students have repeatedly won awards at International Olympic mathematics or physics contests. However, the culturally derived norms and societal expectations have also left some negative effects on a child's way of learning. There has been a general concern among educators with children's inability to do well in problem-solving tasks, to think critically, and to be engaged in difficult learning tasks, as compared to their international counterparts (e. g., Japanese children). Despite parents' and teachers' high expectations and the huge amount of worksheets and homework, many students have developed negative attitudes toward school and learning. It is worthwhile to explore how educators and society can deal with the paradoxical experience in China's only children of being the happiest but the least cheerful.

Dealing with the paradoxical phenomenon in the only child is more of an educational issue in urban settings than in China's rural areas. There is still an imbalance between urban and rural areas in terms of accepting the idea of having one child per family ("China's Illegal Families" 1996). The one-child program has been fairly successful in urban areas because many young people claim that it makes no difference whether their only child is a boy or a girl. In contrast, among rural farmers, especially in some economically underdeveloped areas such as Gansu, boys are still preferred to girls. It will take a long time to change their attitudes. Some families still keep trying to have a boy, even if they have already had two or three children. As a result, more and more irrigated land has disappeared in order to build new houses and create new farms. "Yet the desire for a son outweighs concern for the future of the village or, indeed, the nation" ("China's Illegal Families" 1996, 36). China's population control in rural areas is still facing tough challenges. Moreover, population is seen primarily as a familial rather than an environmental issue.

Environmental Education in China

Evidence shows that the government population and environmental policies are not as effective as they were intended to be. Educators (Zhang 1996; Yu 1996) have argued that population and environmental issues should also be addressed by way of enhancing the entire population's education level and promoting EE in elementary and secondary schools. One of the lessons that needs to be drawn from China's population overgrowth during the 1950s and 1960s is that there was no environmental awareness in the public and no EE in schools. Since the 1970s, EE has been more and more emphasized by educational policymakers.

In China, elementary and secondary schools have adopted an "integrated" approach to EE, which means that environmental topics are integrated into subject areas. Only a few schools have EE courses for students to take as electives. In addition to these courses, students acquire knowledge about the environment through a variety of extracurricular activities and after-school programs, such as attending environment-theme seminars and shows; watching movies; conducting research on the home, school, or factory environment; and participating in Earth Day, Tree-Planting Day, and summer environmental camps.

However, China's EE still faces formidable challenges. Some administrators and teachers are reluctant to integrate environmental topics into their courses because they believe that students' knowledge about environmental issues does not affect their academic achievement or the school's reputation. Things are much worse in schools in rural areas, where more than 80 percent of the people live. There is no EE at all in some schools, and teachers who teach the course have not received any formal training in EE. These schools also lack resources, such as books, audio- and videotapes, and instruments, to deliver a high-quality EE curriculum. Yu (1996) has warned that a failure in EE among elementary and secondary school students now and reluctance to promote the entire population's awareness of environmental issues would lead to serious consequences in China's already damaged environment. The lesson of population overgrowth during the 1950s and 1960s must be learned! China's present population explosion is due in large part to the lack of population education and public awareness of environmental issues. If China does not educate its students to protect the environment now—before it is too late—living in "harmony with nature" will only be the legacy of the distant past.

References

Altman, I. & M.M. Chemers. 1980. *Culture and Environment.* Belmont, CA: Brooks/Cole Publishing Co.

Chai, Ch'u & Winberg Chai. 1965. *A Treasury of Chinese Literature.* New York: Van Rees Press.

Cheng, Chung-ying. 1986. "On the Environmental Ethics of the Tao and the Ch'i." *Environmental Ethics 8* :4, 351-70.

China's Illegal Families. 1996. *The Economist* (February) 36.

Cotterell, A. & D. Morgan. 1975. *China's Civilization.* New York: Praeger.

Heath, S. B. 1982. *Ways With Words.* Cambridge, UK: Cambridge University Press.

Huff, T. E. 1993. *The Rise of Early Modern Science: Islam, China, and the West.* Cambridge, NY: Cambridge University Press.

Leys, Simon. 1996. Review of J.F. Billeter, *One More Art: The Chinese Art of Writing. The New York Review* (April 18) 28-31.

Li, Xiao-ming. 1996. *"Good Writing" in Cross-Cultural Context.* Albany, NY: State University of New York Press.

Wertsch, J. V. 1993. *Voices of the Mind.* Cambridge, MA: Harvard University Press.

Yu, Huiyin. 1996. "Bixu Jiaqiang Zhongxiaoxue Huanjin Jiaoyu" ("Promoting Environmental Education in Secondary and Elementary Schools"). *People's Education 6,* 30-31,

Zhang, Jianru. 1996. "Wuomen Zhi You Yi Ge Diqiu" ("We Have Only One Earth"). *People's Education 6,* 27-30.

Chapter 7

Environmental Education in Italy: Thinking Globally and Acting Locally

Raffaella Semeraro, Michela Galzigna, Lucia Mason,
Marina Santi, and Paolo Sorzio
University of Padua, Italy

Environmental issues occupy a top place among the many world problems, and EE has been receiving increased attention in recent years at the international, continental, and national levels. Principles, operational frameworks, and viewpoints on promoting and developing EE in Italy must, therefore, be seen against the background of international organizations' activities and continental (in our case, European) policies.

Environmental Education in Italy: The Political Framework

The first aim of EE in Italy, as elsewhere, should be to make people aware of the importance of environmental problems and involve them in the elaboration of a plan for the future of the planet. In this way, new mentalities and attitudes will be created, starting from a hierarchy of priorities, as well as problem-solving procedures, that can be established and applied not only to individual needs but also to collective social problems that affect all nations.

We will leave to one side references to general state policies on environmental emergencies and focus more specifically on those concerning EE. Although this problem has been acknowledged by the Italian government, Italy has not as yet produced any clear political guidelines on intervention programs tailor-made for young people and adults in the field of EE and training. Agreement protocols (i.e., outlines of agreements) between the Ministry of Education and the Ministry of Environment were exchanged in 1987, 1991, and 1994. Moreover, a number of circulars (the latest in 1991) and "A Ten-Year Plan for the Environment" were sent out nationwide by the Ministry of Education (1993). Although they highlighted the need for joint interventions to educate young people about the environment, national plans for the organization and implementation of goal-directed and long-lasting interventions in this field are still far from being realized.

Since the beginning of the 1990s, a number of programs, financed by both ministries, have been under consideration at the Centro Europeo dell'Educazione in Frascati, near Rome, and at the Consiglio Nazionale dell Ricerche in Rome, with the aim of exploring possible contact points between education (teaching/learning processes) and environmental concerns.

Following the Ministry of Environment's implementation in recent years of a program for the protection of natural areas (preservation and creation of natural parks), the Ministry of Education has tended to identify EE with setting up particular learning situations in schools that orient students toward this kind of environmental protection. This perspective, however, focuses on only a small part of the wide range of problems with which EE should be concerned.

An Outline of Research and Didactics in Italian Universities

Italian universities are rich in faculties and research departments where environmental problems are dealt with comprehensively, especially in regard to the biological and physical sciences, laboratory research, and the application of new technologies.

Since 1990, some degree courses in EE have been established nationwide in Italy. While the courses are a recent innovation, they do not envisage university training in the field of EE, nor do they plan for modalities to deal with problems of the social and anthropic world. Therefore, fundamental references to the problematic relationship between human societies and the environment are lacking, as are the scientific orientations needed to inform and train both young and adult populations.

Recently, the teaching of EE was introduced into the just-reformed four-year degree program in educational sciences. We hope that this discipline will be part of the future degree courses now being set up to give university training to all teachers in Italy, from nursery to high school levels, mandated by Law 341 of 1991, which envisages a sweeping reform of university organization concerning the initial training of all school teachers.

Although within the Departments of Educational Sciences set up at the Faculties of Magistero (now being changed into Training Sciences Faculties), research groups are concerned with EE (especially at the Universities of Bologna, Florence, and Padua), and despite the Italian Research Center in Environmental Education in Parma (central Italy), there is as yet no widespread university didactics training students to face such problems.

At the University of Padua, Semeraro is trying to set up a postgraduate school open to all graduates, with the aim of training programmers, administrators, and operators in EE. The new professional training furnished by such a school should be expressed in the following areas:

1. in-school educational area (people responsible for the administrative and didactic management of elementary, middle, and high school teachers, as well as teachers in initial and inservice training);
2. the out-of-school educational area (people operating in the fields of continuing education and in training agencies nationwide)
3. the business field (from material to cultural production) in the relationship between industry and ecology
4. the communication and mass-media field (communication and information experts, cultural operators)
5. the administrative-political area (administrators of training interventions on the environment in towns, regions, and in state policy)
6. the social area (analysts, programmers, and those planning interventions on demographic aspects, migrations, and labor mobility)

The specialties envisaged by the school in relation to these areas are (1) EE, school, and continuing training (areas 1 and 2); (2) EE, science, and technology (area 3); (3) EE, languages, and communication sciences (area 4); (4) EE, from political-economic and human-science perspectives (area 5); and (5) EE, from historical-geographic and social-science perspectives (area 6).

By referring to this initiative, now taking shape at the University of Padua in collaboration with UNU (United Nations Universities), we wish to underscore the importance of orienting the university community (and all the faculties and research fields featured by the universities) toward a connection between disciplinary and technological knowledge and the effects of their application on the environment.

The model proposed here is that of diffusing knowledge and learning and addressing the ensuing need for interdisciplinarity in scientific research and for interdependence and interaction among the different groups that make up a society (politicians, administrators, scientists, technicians, communication and information experts, institutional and organizational leaders, industrialists, workers, teachers, parents, young adults, the elderly, etc.).

It is therefore necessary to adopt an EE model that plans for a distribution of tasks and responsibilities between institutional and non institutional groups in order to acquire widespread training oriented toward alerting the population to the problem of protecting natural and cultural resources and to the problems everyone must face following the adoption of a sustainable development model. With respect to EE, we must talk about an integrated system of education and training.

In regard to Italy and its public organization, this model has been explored and is being investigated further. The interventions already carried out in the area of EE can be considered experimental procedures of social responsibility distributed throughout this field.

The Development of Environmental Education

A typical Italian phenomenon is regional and local decentralization at the political, administrative, cultural, and social level. Since the 1970s, there has been a unification process among national, regional, and local policies involving all sectors of public life. Hence, choices and checks on social life involve not only the state but regional and local bodies as well.

This aspect of Italian public life is not without tension in the management of state power and control versus regional and local power. However, it has encouraged many EE initiatives outside the school to stimulate young students and their teachers to project and implement research studies, as well as cultural and social activities (exhibitions, debates, visits to natural and/or urban areas). Over the last 10 years, this has helped to focus attention on the environment through EE. Leaving aside the school system, there is a range of interventions in the fields of cultural activities and adult populations' continuing education that have raised people's consciousness of environmental issues.

Out-of-School Environmental Education

There are various initiatives in the fields of information, communication, and education specifically meant for young adults and aimed at promoting new knowledge and developing new attitudes toward environmental problems and emergencies. Some initiatives are designed for the school; others are aimed at selected groups of citizens, such as workers' associations in firms, craftsmen, and ordinary people in different town districts. These activities are generally promoted by the following groups:

• regional and/or local bodies: branches of the Council Office for Ecology, Culture, and Education acting as agencies in charge of the central government's interventions in regions and towns
• institutions and cultural centers: organized groups, often not-for-profit, with the aim of awakening people, especially adults, to socially relevant issues
• environmental associations: Italia Nostra, the League for the Environment, and the Italian branches of the WWF and Greenpeace both for the school and the general population

Environmental Education in School

Regional Bodies

In Italy, there are regional Research, Experimentation, and Educational in-Training Institutes with the task of coordinating the modernization of the school system but without power in regard to legislation for the reform of elementary, middle, and higher schooling. Many of these institutes (especially in some regions, i.e., Emilia Romagna, Lombardy, Piedmont, Tuscany, and Veneto) have coordinated projects between schools to develop teaching/learning activities in EE.

For two years, the University of Florence (Prof. Paolo Orefice) and the University of Padua (Semeraro) have set up a national coordination among the regional institutes so that all intervention programs follow established guidelines and are consistent with European and international orientations. This has involved national initiatives in didactic experiments and inservice teacher training.

Provincial Education Offices

These are government offices for elementary, middle, and high schools, and they are in charge of school activities at the provincial level (subregional conglomerates grouping a number of towns). These offices link the initiatives of individual schools with out-of-school cultural activities.

Individual Schools and Networks of Several Schools

Individual schools, especially in central and northern Italy, have taken up the EE issue and have developed it, especially in the compulsory years in particular (11- to 14-year-olds), whereas high schools (14- to 19-year-olds) seem not to have been involved to the same extent. Students have experienced environmental phenomena through field research on the physical, natural, and human environment in rural and urban areas.

Research at the University of Padua

For 20 years, Semeraro has been working for the diffusion of EE in Italy by applying new epistemological principles (i.e., contextuality) within the paradigm of complexity to a dynamic concept concerning both social (cultural systems) and individual knowledge (personal conceptual structures). Many theories and policies concerning cultural diffusion processes have been debated both in school and in social life. Particular interest has been devoted to the change of teaching methods and to knowledge construction processes. Aspects of these themes are being investigated by the other four members of the research group.

Between 1978 and 1981, Semeraro was the scientific director of a pilot project, under the auspices of UNESCO, the Italian Ministry of Education, and the Giorgio Cini Foundation in Venice. In this project, the transfer of an interdisciplinary model already applied to scientific research was explored (Apostel et al. 1983; D'Hainaut 1986; Semeraro 1982, 1983, 1988) and has been introduced to the planning of middle school education, bearing in mind the development of EE. In Italy, a group of university teachers (humanists, scientists, artists, and communication and information experts) have followed the final three years of compulsory schooling, the teachers of preadolescent pupils (ages 11–14) in Venice (pupils from the middle classes), Spinea (pupils from farming and small-contractor families), and Marghera (pupils from working-class families). The aim was to see whether the teachers' new didactic interventions, (which proceeded from the application of an interdisciplinary mode to planning of teaching/learning contexts according to cooperative practices and procedures between teachers) alerted the pupils to the problems of urban environments irrespective of their background. The results suggest that:

1. EE needs interdisciplinary perspectives to be applied to scientific research and curriculum planning.
2. Cooperation between teachers, although difficult at first, produces a greater number of communicative exchanges between the teachers themselves and between themselves and the students.

3. Research carried out on the environment, either in or out of school, must change from information about environmental issues and emergencies to problem recognition so that students may develop knowledge and action strategies based on problem solving. Different environments (physical, natural, and social) are then experienced as knowledge sources for human beings as well as areas in which the appropriateness of solutions is continually checked in the process of integration, survival, and development.

4. The interactions among different events present in environmental contexts (according to different phases of organization, disorganization, and transformation) is reflected in the dynamics of people's thinking, communication, and behavior processes. This calls for the need to rediscover and value a new alliance among people, society, and environment according to biocentric principles and orientations of sustainable development.

Over the last ten years, Semeraro has conducted research at the university level and interventions that investigated these approaches. In recent years, collaboration has been established with ENEA (Italian Body for New Technologies, Energy, and Environment) which is showing interest in EE and in the newly emerging environmental professions.

Experiments carried out, school initiatives, and social interventions are all within the spirit of the orientations stated in the *Principles of the Declaration on the Environment and Development* and in *Agenda 21*, passed by the Rio de Janeiro UN Conference. Further, Chapter 36 of the same Agenda (promoting Education and Public Awareness and Training) underscores the need to orient education again in the light of a better human/environment relationship. Emphasized also is the need to create new mentalities and awareness not only in each individual but in all people.

A real "revolution" of thought and behaviors (Brown 1992) must be promoted, the meaning of production and work processes (European Commission 1994; Timberlake 1992) must be changed, new networks of cooperation among institutions, governmental organizations, and NGOs must be instituted at the local and global levels (Kimball 1992). Above all, we must work together to establish priorities (Green 1992) around which to mobilize human activities.

Knowledge Construction as Revision of Pre-existing Conceptions

From the first level of schooling, EE is important because it enables children—future citizens—to construct knowledge, skills, values, and attitudes that allow them to become effectively engaged in the best development of our society (Semeraro 1988). Children have been exposed to many ideas and opinions concerning the ecological fate of the planet and the species (human and nonhuman). When their conceptions of environmental problems are taken into account, they may form the basis for an effective ecological curriculum. EE can be a "model for constructivist teaching" (Klein & Merritt 1994) based on the belief that learners actively construct their own knowledge instead of passively receiving it from the outside.

Constructivist Teaching and Conceptual Change

Children hold images and representations (see Mortari, this volume), through which they interpret and make predictions about the world that affect their reasoning about environmental issues. In the past two decades, research in cognitive and developmental psychology (see McNamee, this volume) and science education has shown that individuals construct personal knowledge structures on the basis of their everyday experience (see Stanisstreet & Boyes, this volume). These structures are often in stark contrast with scientific conceptions (Duit 1994). Knowledge of children's understanding of pollution, for instance, may help educators to create a variety of meaningful classroom experiences regarding pollution, including solid and toxic waste and air, water, and soil pollution (Brody 1990-91).

In a recent study carried out by two members of our research group (Mason & Santi 1994), at the beginning of the implementation of an EE project, a very partial, little articulated concept of pollution in 5th graders who lived in a country village was found. Only one child referred to soil pollution due to the use of chemical fertilizers and pesticides in agriculture. Several children stated that chemically treated fruit and vegetables were better and healthier than those grown naturally since they were not "ill because of a worm inside." These are relevant data if we consider that the children lived in a rural area and knew that farmers had to wear masks to spray several crops with chemicals and that produce could not be picked before a set time had elapsed. Moreover, one child strongly maintained (against the opinion of all other classmates) that manure was not a natural fertilizer since "it is made of dung, and I'm sure dung is dirty and brings diseases." On the contrary, she maintained that "[c]hemical fertilizers are safer because they don't bring diseases."

What emerges is that students should be involved in a process of conceptual change rather than in a mere increase of information in order to revise the entrenched beliefs on which their representations of phenomena are based. From the cognitive point of view, an appropriate EE intervention should be aimed at the revision of alternative conceptions about environmental phenomena.

The Triple Alliance

Environmental education is a field in which we cannot be successful if, as educators, we do not take into account what has been defined as "the triple alliance," i.e., the involvement of motivational, cognitive, and metacognitive aspects (Short & Weissberg-Benchell 1989). The critical role played by motivation, that is, the willingness to learn or act in general and in a specific topic and task, to spend appropriate energies and commitment in learning and acting, is well established. In EE, children's experiences and emotions (see McNamee, this volume) can be powerful sources of personal involvement that, in turn, help to create meaningful contexts in which to ground new knowledge.

The role played by cognitive aspects is even more well established. They are fundamental in constructing declarative and procedural knowledge concepts, principles, problem-solving strategies, transfer skills, etc. In EE, higher-order thinking and reasoning skills can be stimulated and sustained through the very characteristic of having disciplinary content about the environment, which entails sophisticated cognitive procedures, such as establishing relationships, connections, and interdependencies among elements, phenomena, and systems.

The role played by metacognitive aspects has been acknowledged more recently. The term *metacognition* has a multidimensional character and is often used in different ways. In an EE project, it can be understood mainly as the inner awareness of, or ability to reflect on, one's own as well as others' beliefs, theories, values, actions. Further, it deals with reflections about knowledge as well as about oneself as a person who thinks, knows, and acts. Here, we explicitly refer to the meaning expressed by Kuhn, Amsel & O'Loughlin (1988), who made a distinction between thinking *about* and thinking *with* a theory. To be able to think *about* a conception on an environmental topic leads to metaconceptual awareness of its epistemological, ontological, and ethical values or limitations. This creates a fruitful ground for revising one's knowledge about the environment.

Discussion-Based Learning About Environmental Issues

A developing research trend characterizes learning in general, and the development of specific understanding, as a social rather than an individual process. In the perspective developed by social constructivism (Resnick, Levine & Teasley 1991; Rogoff & Lave 1984; Wertsch 1985), knowledge growth and change are the product of personal interactions in social contexts and of appropriation of this socially constructed knowledge. Social interaction in a classroom, considered as a natural social context, has been recognized for its cognitive potential for thinking/learning. The focus of many recent studies on peer interactions in classrooms is on collaborative learning as a means of fostering higher-level thinking and reasoning processes. Indeed, looking for reasons, advancing justifications, trying to explain something, opposing rebuttals, hypothesizing solutions, evaluating evidence, and considering alternative positions are all activated in classroom discussions. By constructing justifications of an idea, a learner often integrates and elaborates knowledge at more advanced levels (Brown 1995; Brown et al. 1993).

These aspects of this "triple alliance" are particularly emphasized in collaborative learning based on group discussion. This can stimulate the motivations of learners, giving them the chance to bring their experience into the class, to express their ideas, which are not immediately judged as right or wrong but respected, and to discuss relevant current events and problems with others. The social context of learning activities gives the mental effort shared among the participants greater support at the emotional level by reducing their anxiety about being evaluated on their own cognitive functioning. The higher-order cognitive procedures needed to construct new knowledge about the environment are better stimulated in situations that require reasoning and critical thinking.

Finally, the metacognitive dimension is involved in that reflections on the plausibility and reliability of one's own representations are continuously stimulated, as well as the consciousness of the need (essential in EE) for coordinating theory and practice, that is, consistency among our values, beliefs, and conceptions and our everyday individual and social actions.

Conclusions

In conclusion, the efficacy of efforts in EE can be evaluated insofar as they produce relevant changes in individual and social actions. Environmental education should result in socially important actions expressing individual commitment toward the solution of ecological problems dealt with during the educational intervention (Posch 1993). The conceptual changes induced by new learning should be accompanied by a change in attitudes, motivations, and actions in regard to the environment. Interventions in EE should connect what people do, do not do, and should do, with what people know, do not know, and should know. Up till now, environmental problems are faced by encouraging/discouraging some actions on the basis of a certain corpus of knowledge, mostly from the physical and biological sciences.

A large part of the educated population, though receiving relevant information on environmental problems from the mass media, acts inappropriately. Many who declare their environmentalist faith do things incompatible with it, even without recognizing inconsistencies. That what is being done is based on ideas sustaining or disclosing systems of conceptions and beliefs seems to be lacking. Therefore, EE should try to bind together theories, values, and actions so that knowledge about the world manifests itself also as an action on the world, and every positive action makes explicit an evaluated and chosen project. In other words, it means educating *through*, *on*, and *about* the environment.

From the perspective of EE, a key element in changing attitudes and behaviors toward the environment is giving learners the opportunity to construct relevant knowledge that is powerful when translated into appropriate actions and positive values. Instead of turning to "the experts," future citizens will have information and know how to use it. By learning to reason and argue with others on environmental problems since elementary school, they will be more likely to be interested in taking part in "communities of reasoned discourse on public issues." Thinking globally and acting locally also means helping children to appropriate "a culture of reason, analysis, and reflection, based on certain shared knowledge. Realizing this vision will require a civic consciousness that goes beyond the individualistic one of current classroom learning models and draw on models of shared intellectual functioning such as we see in our best work environments" (Resnick 1987, 19).

References

Apostel, L. et al. 1983. *Interdisciplinarité et Sciences Humaines.* Paris: UNESCO (In French).

Brody, M.J. 1990-1991. "Understanding of Pollution Among 4th, 8th, and 11th Grade Students." *Journal of Environmental Education 22*:2, 24-33.

Brown, A.L. 1995. "The Advancement of Learning." *Educational Researcher 23*:8, 4-12.

_____. et al. 1993. "Distributed Expertise in the Classroom." In G. Salomon, ed. *Distributed Cognitions. Psychological and Education Considerations.* New York: Cambridge University Press, 188-228.

_____. 1992. "An Environmental Revolution." In UNCED (The United Nations Conference on Environment and Development), *Earth Summit '92.* London: The Regency Press Corporation, 18-21.

D'Hainaut, L. 1986. *L'interdisciplinarité dans l'Enseignement Général* Paris: UNESCO.

Duit, R. 1994. Conceptual Change in Science Education. Paper presented at the "Symposium on Conceptual Change," Jena, Germany.

European Commission. 1993. *Towards Sustainability. A European Programme of Policy and Action in Relation to the Environment and Sustainable Development.* Brussels.

European Commission. 1994. "Crescita, Competivita, Occupazione Le Sfide e le vie per Entrare nel XXI Secolo." Brussels, Belgium.

Green, R. 1992. "Priorities for the Future." In UNCED (The United Nations Conference on Environment and Development) *Earth Summit '92.* London: The Regency Press Corporation, 32-6.

Kimball, L.A. 1992. The Institutions Debate. In UNCED (The United Nations Conference on Environment and Development,) *Earth Summit '92.* London: The Regency Press Corporation, 38-41.

Klein, E.S., & E. Merritt. 1994. "Environmental Education as a Model for Constructivist Teaching." *Journal of Environmental Education 25*:3, 14-21.

Kuhn, D., E. Amsel, & M. O'Loughlin. 1988. *The Development of Scientific Thinking Skills.* San Diego, CA: Academic Press.

Mason, L., & M Santi, 1994. "Argument Structure, and Metacognition in Constructing Shared Knowledge at School." Paper presented at the Annual Meeting of the American Educational Research Association,

New Orleans, LA. (April).

McCormick, C. B., G. Miller & M. Pressley, eds. 1989. *Cognitive Strategy Research: From Basic Research to Educational Applications.* New York: Springer-Verlag.

Posch, P. 1993. "Research Issues in Environmental Education." *Studies in Science Education 21*, 21-48.

Resnick, L.B., 1987. "Learning in School and Out." *Educational Researcher 16*:9, 13-20.

Resnick, L.B., Levine, J., & S.D. Teasley, eds. 1991. *Perspectives on Socially Shared Cognition.* Washington, DC: American Psychological Association.

Rogoff, B. & J. Lave, eds. 1984. *Everyday Cognition. Its Development in Social Contexts.* Cambridge, MA: Harvard University Press.

Semeraro, R. 1982. L'interdisciplinarita nell'Insegnamento. Firenze: Le Monnier.

_____. 1983. *Dinamica della Conoscenza e Comunicazione Interdisciplinare.* Rome: MPI-Isituto Della Enciclopedia Italiana.

_____. 1988. *Educazione Ambientale, Ecologia, Istruzione.* Milano: Franco Angeli.

Short, E.J., & J.A. Weissberg-Benchell. 1989. "The Triple Alliance for Learning: Cognition, Metacognition, and Motivation." In McCormick, Miller & Pressley (above), 38-82.

Timberlake, L. 1992. "Changing Business Attitudes." In UNCED (The United Nations Conference on Environment an Development), *Earth Summit '92.* London: The Regency Press Corporation, 29-31.

WCED (World Commission on Environment and Development). 1987. *Our Common Future.*

Wertsch, J.V. ed. 1985. *Culture, Communication and Cognition: Vygoskian Perspectives.* New York: Cambridge University Press.

Part II

Eco-politics and Environmental Ethics

Chapter 8

EXPO 2000 in Hannover: Eco-politics in Germany and the Universal World Exposition

Frank E. Puin
Social and Ecological Technology Research Center
Hannover, Germany
and
Fachhochschule Fulda
Fulda, Germany

The city of Hannover is typical of northern Germany. The state capital of Lower Saxony, Hannover has an interesting historical relationship with Great Britain, which was ruled in the 18th and early 19th centuries by the Hannoverian line of British monarchs. Today, Hannover is part of the upper portion of a European high-tech industrial "banana" that stretches from Glasgow in the north to Barcelona in the south. The industries in this area are highly competitive—not only in Europe but in America and Asia as well. Within this context we can examine the forthcoming EXPO 2000 in Hannover from eco-political and socio-ecological perspectives.

The Political Backdrop

Since German reunification and the collapse of the Soviet Union, Hannover is again situated in the center of Germany and of Western Europe. Characterized as the "central station" of the Federal Republic, Hannoverians see themselves as the heart of Germany and of Europe. "The third millennium begins in Hannover," they say, perhaps without considering the lingering distaste of those who recall Germany's recent role on the world stage.

Since World War II, the city of Hannover and the state of Lower Saxony have been dominated by the Social Democratic Party (SDP), which was founded there in 1945. After the "cultural revolution" in Germany in the late 1960s and the crisis in traditional engineering and automobile manufacturing in the 1970s, the SDP gradually lost its political hegemony and formed a "big coalition" with the conservative Christian Democratic Union (CDU) and Free Liberal Democratic Party (FDP). In the 1980s, the Green Party emerged as the only opposition to the coalition.

In the political milieu of the big coalition, the idea of a huge event in Hannover—a universal EXPO—was born and approved by the Federal government, led by the conservative chancellor Helmut Kohl. However, the EXPO 2000 momentum did not stanch the decline of the big coalition in Lower Saxony. In the early 1990s, the SDP and the Greens won elections and formed a "Red-Green" coalition. This eco-social majority saw the prospects of EXPO 2000 in very different terms from its originators and also from its current planners—namely, the federal government in league with international organizations and various local groups. The new majority, assisted by a neutral citizens' information bureau, organized a plebiscite in 1992 which the pro-EXPOs won by a narrow margin (51.5 percent).

Two years later, the Red-Green coalition disintegrated on the question of financing the urban part of the city's infrastructure needed to launch EXPO. But the alliance was not completely destroyed. On the state level, the SDP was able to govern alone after the 1994 elections; the new government ordered funding for the EXPO event to be drawn from the state's financial reserves, but these are insufficient. To make matters worse, the social fabric locally has been threatened by growing unemployment (which has reached 13 percent), a city budget deficit of about 200 million DM in 1994, and "chaos days" (youth riots). The saving target in the pre-EXPO phase amounts to about 400 million DM of the city budget and about 2.5 billion DM in the State budget. The losing anti-EXPOs believe that time will show the inherent contradictions of the EXPO motto "Man, Nature, and Technology" and will also reveal the negative effects of such an enormous event on the local ecology and economy.

What impact would the enormous 153-day EXPO event now being planned for Hannover in the year 2000 have on the regional ecology and economy? Consider the following points:

- If all infrastructure projects to handle 40 million visitors (an estimated 300,000 a day for 153 days) are completed, Hannover's ecology will necessarily be damaged. If these vast numbers of people really do invade the city—doubling its population for half a year—emissions from waste and traffic and consumption of energy and water will damage the environment and the quality of life for residents.

- On the other hand, the economy will be damaged if these 40 million people do not come and spend at least $100 a day! (The break-even point has been set at about 4 trillion DM—about 70 percent of direct return on investment, 30 percent of indirect.) The currently stagnating local economy will be further damaged if it must also bear the costs of an overdeveloped infrastructure.

The choice to hold a universal EXPO—instead of a smaller, specialized one—was a decision against sustainable financial and economic development of the city and the region. The economic winners will be global corporations, foreign construction workers, and strategically located restaurants and hotels. The rest of Hannover—including its local communities—will be among the losers.

The EXPO organizing groups and local politicians in Hannover are torn between feelings of omnipotence and provinciality. This is regrettable, because the city has a tradition of staging some of the world's great industrial fairs. But the ambition that led to the idea of a universal EXPO was fueled by a popular position in German political discourse that seeks to resolve social and ecological problems without reducing economic productivity and within the framework of a democratic society. This is the result of a failure among Germans to face their own history honestly, which is compounded by a general lack of understanding of economics, eco-politics, and philosophy.

Losing the War but Winning the Peace

Germany, with other countries around the world, has just commemorated the 50th anniversary of the end of the Holocaust. Few Germans remarked, as did former President R. V. Weizäcker, that 50 years ago the "liberation of Germany from the Nazi regime" also occurred. Right-wing parties and many conservatives recalled instead the defeat of Germany and the beginning of another tragedy—the banishment of Germans from their eastern homes, the prolongation of totalitarianism in eastern Europe and Russia, and the new national socialist movements in other parts of the world. Some young people sprayed graffiti, such as "no tears for krauts," but most Germans seem to hold the belief that "we lost the war but won the peace."

EXPO 2000 may be seen as the celebration of this victory of the peace. The idea is to present a modern, conscious, peacefully reunified Germany, looking forward optimistically and working hard to solve world problems, especially the conflicts between ecology and economics. But this is the bright side of a country that does not recognize its obligation to learn from its own history. It is instructive to consider what is not planned for EXPO 2000: a history of European fascism, for example. Nor are Auschwitz and Bergen-Belsen, which are located near Hannover, mentioned in the EXPO concept. There will be no exhibits on the Holocaust technology of death or on the German military-industrial complex—nor any ideas on how to prevent fascist, racist technocracy in the future. Western Germans since 1968 have learned techniques of civil resistance. Nonviolent protest has become one of the most important instruments of the modern ecological movement in Germany. But "social defense without violence" is not always successful, as one sees in the case of EXPO.

Spirit, Nature, and the Smog of Modern Enlightenment

In the 1980s, philosophical postmodernism seemed to overrun ideological positions, such as socialism, ecologism, liberalism, nationalism, Thatcherism, capitalism, Catholicism, and Islamism. Germany has organized many congresses to analyze philosophical relations among "man," nature, and technology. Germans contemplated different ethics, while the emergence of the New World Order proceeded without them. One congress that took place in Hannover discussed abstract interrelations of spirit and nature, while the Club of Rome remodeled its pattern of ecological disaster. The idea of EXPO 2000 was created in this atmosphere of New Age expectations. The event, it would seem, should encompass the great transformation in modern society from declining industries to service technologies, know-how, technological progress in the biological sciences, and the global expansion of "Silicon Valley."

The beginning of a new long wave of technological development needed philosophical ideas. "Back to Kant!" was the rallying cry of the more knowledgeable EXPO promoters. One of Kant's messages is sustainability: the slow clearing of reason, honoring "the stars in heaven above me and the moral law within me" and working out the suppositions of a good and peaceful life for all people. His recommendations for realizing everlasting peace for all nations were illumination and rationality, contentment within the social framework, progress in the sciences, and the rejection of political decision-making by irrational masses. An antidemocrat, perhaps Kant anticipated the mass phenomenon of fascism. In this sense, Albert Einstein's

"Letter to the American President" written in 1939, shortly after the opening of the New York World's Fair, to encourage the efforts of building the nuclear bomb, seems to have been a necessary act for the civilized world. The German presentation at the 1939 World's Fair was as audacious as philosophical fascism was aggressive. In 1954, after the release of nuclear power, Einstein said the relation of man to nature and to technology had changed fundamentally. The principle of self-destruction had become global.

The final historical point of technological development is the atomic industry. Its "peaceful" branch was pushed forcefully by German governments—and without hard resistance until the protests of wine farmers in the Rhine area and of physicist dissidents, such as Robert Jungk. The political ecological movement in Germany owes much to both Einstein and Jungk, as does German industry when it distinguishes itself as the leader of worldwide eco-technology.

But the planned EXPO 2000 will not address world change as a result of the atomic bomb and the "peaceful" circle of nuclear power and waste production. The same logic will suppress the Chernobyl accident in the Russian pavilion and the nuclear arsenals in North Korea, Iraq, and other countries. And Germany, France, and Great Britain will be silent about their programs for the transportation and disposal of nuclear waste.

The risk society has a philosophy: total quality management (TQM) and zero fault. But nuclear technology cannot allow for the old technique of learning by doing; one extraordinary event can determine self-destruction versus development. The new technologies of the modern world need to employ Kantian principles of everlasting peace. EXPO 2000 bears its own risk for self-destruction—without the possibility for "learning by doing" or employing the successful strategy of "muddling through."

The Balance of Nature and the Technology of Intervention

Since Adam Smith's *Theory of Moral Sentiments* (1759) and The *Wealth of Nations* (1776), the self-regulating "invisible hand" has been the paradigm of modern economics and of a modern psychology of needs. However, this paradigm lacks the social boundaries of a moral economy as conceived by Smith.

The modern capitalist brain trust was organized within the Club of Rome (which began its work in Hannover in 1970, then moved to MIT in the United States). With the study *The Limits of Growth* (1972), it recognized the end of self-regulating economies of growth. The Green movement and civil-rights groups gave up on strategic and general planning—at least on the part of the government. If one agrees that small is beautiful, one needs central planning centers only in infrastructural branches of supply for the people. Hannah Arendt was skeptical of supranational intervention or even of demonstrations, world meetings, and mass assemblies, except in the case of ethnic genocide, as in fascist Germany. She preferred local neighborhood intervention.

Germany has had a traditional emphasis on federalism, regionalism, small firms, handicrafts, and neighborhoods. Real power is also held by labor union functionaries, by middle managers in centralized bureaucracies, and by energy and chemical industry copartnerships between labor and capital. These are the groups that would realize most of the benefits of EXPO 2000.

Two main groups are taking part in the planning for EXPO 2000: managers and politicians. The management group, which seems to have been disappointed by market constraints, is now looking for "public effects"—in urban development, architecture, exposition, new media technology, and a secure working place. The politicians are calling for more private risk and entrepreneurship; in effect, they want to be political decision makers without taking other than political risks. This public-private partnership in a 10 billion DM project like EXPO 2000 promotes a mixed economy between private and public sectors. The risks and losses will be socialized; the benefits, privatized. This cannot be fair to the rest of the people unless a broad majority accepts the project. But under current conditions, accepting EXPO 2000 would require Hannoverians to turn over their environment and economy to the globalization efforts of a worldwide network made up essentially of strangers to the region. The anti-EXPOs hope to cooperate with NGOs around the world. Skeptical of governmental representation, they are seeking NGO help in migration and market development for small and adapted technology.

If we follow A. O. Hirshman's law of "ups and downs" in private/public activities, EXPO 2000 is a play with changed parts: Industrial hardliners play the role of ecologists; ecologists play the role of sustainable economists. Eco-groups waiver between disappointment and engagement; that is, they see opportunities to contribute to the content but cannot join the private/public economic conditions.

What is the content of EXPO 2000? The published concept describes "houses" (i.e., pavilions and national expositions) in which the EXPO motto is interpreted through elements, sentiments, pictures, ideas, participation, and visions—combined in a "park of sense." Exhibitions are planned for areas in health and nutrition, environment, development, living, working, communication, information, mobility, and education.

The city of Hannover wants to show itself as an urban garden, a social living area, and a modern city in the year 2001. The surrounding region and the State of Lower Saxony want to show off a "coastal landscape" with 20 more projects of ecological living and working. If this universal approach shows the mainstream Western lifestyle (even in its expensive environmental variation of European urban centers), it will compete with alternative ideas of sustainable living in less developed areas of the world. While the German eco-movement is fighting to change Western patterns of consumption and production, the organizers of EXPO are trying to modernize existing structures.

In addition, EXPO seems to be a replacement of real activity. Solar-energy power production is lost to California, and the heat and power-plant industry is undeveloped, though it was technologically optimized in Hannover. Until now, the "ecos" have expressed their frustration by resignation. They don't want to be caught up in a misunderstanding of some good ecological issues in EXPO 2000 and at the same time to be involved in misleading messages of the whole event.

The ecological movement in Germany and other Western countries must accept the reality of a slow economic growth that affects the ecosystems in different ways. Compare, for example, these facts about the most environmentally troubling industries in the developed nations of the world.

- Germany is a world offender in the production of chlorine chemicals in nerve poison and of dioxin. The next two disproportionately growing industries are energy and aluminum; the former is based on nuclear power, and the latter is the greatest energy user in manufacturing—with highly toxic byproducts.

- Energy production is the only ecologically problematic industry in the United States whose growth is parallel to the growth of the Gross National Product (GNP). Heating systems in American housing are an ecological scandal. Will the American pavilion in Hannover show intelligent solutions to this problem?

In designing EXPO pavilions under the "Man, Nature, and Technology" motto, what can be expected from poor, young nations such as Kurdistan, Eritrea, Slovenia, or Algeria? What can be expected from the "tiger nations" of Southeast Asia or from the population giants of China, India, Pakistan, Brazil, and Indonesia?

The Clash of Culture, Nature, and Social Technology

Yet another reason for concern about the EXPO concept is its potential for attracting violence. With regional conflicts more dangerous now than during the Cold War, the next wars will happen at the borders of different cultures. Indeed, people carry their frontiers within themselves. EXPO 2000 could be seen as a primary target for terrorism from all parts of the world.

This danger could be diminished if Germany stood for an active policy of peace, development, and economic assistance. On the contrary, the government has cut the budget of the Ministry of Development and has made pragmatic arrangements with governments all over the world, whether or not they are democratic. The international relations of Hannover and Lower Saxony have been somewhat more politically correct, but a world exhibition with pavilions from all national governments needs more than a neutral pact with the rich and mighty.

The German police will likely prepare themselves by adding to their numbers and suppressing opposition groups, as well as those truly interested visitors to EXPO 2000 who are familiar with ecological movements in Germany and elsewhere. In recent years, the conservative German government has not been peace-sustaining or financially generous internationally. About 180 billion DM are transferred annually to the reunited German states in the East. Thus, Germany will not harvest international solidarity because it has not sowed "financial seed money."

Conflicts at EXPO in Hannover might follow this pattern of clashes of civilization:

- conflicts between and within the least developed countries, where people are fighting one another over scarce resources (as Peru and Ecuador did recently)
- conflicts between poorer and richer neighbor nations (such as Iraq and Kuwait)
- conflicts to correct the result of World War II, especially in the former Soviet Union (such as clashes in the Caucasus, among the Baltic nations,

or in the Indonesian island of East Timor)
- conflicts left over from earlier suppressions of cultural identity and national development (such as Kurds, Basques, Berbers, Tatars, Bosnians,
Serbs, and Albanians)
- new conflicts among populations where ethnic and religious groups have
been segregated and suppressed (such as Polynesia)
- predictable conflicts over the local economies of new states or ministates
(such as Palestine/Gaza or Ogoniland in Nigeria)

International communities in Hannover constitute about 10 percent of
the local population and are not well organized. However, the Turkish-
Kurdish and Islamic fundamentalist conflicts will have a local impact. EXPO
2000 will be a cultural event, interpreted by many as an effort to harden the
technical hegemony of Western European and American lifestyles, of their
capitalist economies, and of their prosperity. Any and all forms of conflict
might erupt in response.

Conclusion

We should all reflect on the EXPO event in Hannover from a historical,
ecological, economical, and cultural point of view. Many aspects of this
event have not been explored (city development, architecture, traffic, citizen participation, regional development, changing the costs of production,
and the labor market, to name a few). But those who accept my arguments
will begin the third millennium at home and put off their visit to Hannover
until after the year 2000. The "wind of change" (as a local rock group sang
in 1989) will catch Western nations, just as the Communist states were
caught. Germany and Hannover will not become a voice for the environment until the consequences of an anti-ecological EXPO have been cor-
rected by a political leadership that has adopted authentic environmental
goals and a new culture of ecological economics.

References

Arendt, Hannah. 1964. *Eichman in Jerusalem: ein Bericht von der Banalitat des Bösen*. München: Piper Verlag.

Duve, Freimut. 1975. *Kommunismus ohne Wachstum? Babeuf und der "Club of Rome:" Sechs Interviews mit Freimut Duve und Briefe an ihn.* Reinbek bei Hamburg: Rowohlt.

Huntington, Samuel P. 1996. *The Clash of Civilizations and the Remaking of World Order.* New York: Simon & Schuster.

Jungk, Robert. 1958. *Brighter Than a Thousands Suns: A Personal History of the Atomic Scientists.* New York: Harcourt Brace.

Kant, Immanuel. 1963. *Perpetual Peace.* New York: The Library of Liberal Arts.

Chapter 9

The Bioregional Vision: Prospects and Problems

David W. Shapiro
Ithaca College
Ithaca, New York

T his chapter addresses the benefits, both individual and communal, of the bioregional vision and then suggests the monumental difficulties that would be inherent in attempting to adopt such an approach. Since any profound societal transformation could be said to be difficult to institute, this chapter attempts to assess bioregionalism from the perspective of the relative ease of its adoption in comparison with other environmental approaches.

Bioregionalism

Bioregionalism is an attempt to reestablish a connection between human communities and the larger total biotic community of a given geographical region. The criteria for defining the boundaries of such regions might include similarities of land forms, flora and fauna, and common water-drainage systems. The promise of such a literal and cognitive re-mapping of a place would be to develop within the human community an ecological consciousness that would "respect carrying capacity, engender social justice, use appropriate technology creatively, and allow for a rich interconnection between regionalized cultures" (Aberley 1993, 3). The assumption is that the political boundaries that presently define an individual's sense of place (city/county/state/nation) and the maze of institutions that operate within and between these boundaries would be restructured to reflect shifting sensibilities and economic realities.

Benefits of Re-mapping

Leaving aside for the moment the immense challenge of bringing such a vision to fruition, it is not difficult to imagine the advantages that such a re-mapping could bring about. Perhaps most important would be the perceptual changes that would take place in the minds of individuals and their communities. The natural world might be seen as "a rich and varied collection of living things linked one to another and to their substrate or inorganic environments.... [rather than] merely a stage or backdrop for human initiative and enterprise" (Doughty 1990, 121). Students would spend more time learning about how their local region sustains itself in terms of food supply, water cycles, energy resources, and waste recycling. Local knowledge of the region's flora and fauna, soil types, and weather patterns would assume equal importance with the cultural history of the area. Such foregrounding of the relationships between human communities and the environment could bring about a true ecological consciousness.

The assumption is that once people fully and deeply understand the degree to which they are connected to the larger biotic community, they would endeavor to engage in practices—economic, political, and social—which benefit the long-term quality and sustainability of the total community. Though there is no master plan for what bioregions would look like or how many bioregions there may be, it is fair to conclude that there would be a decentralized approach to all fundamental issues, with a great deal of local community empowerment. Whereas socioeconomic and political changes might have results that look quite different from what presently exists, the notion of decentralization and community control does resonate with both present and past populist movements in the United States.

Difficulty of Changing Minds/Cultures

While it is relatively easy to speculate about the benefits that might accrue from a given cultural transformation, it is quite another thing to adequately address issues in regard to how such a transformation could occur, which approaches to sustainable living might have a greater chance of adoption by the society at large, and what might be the unintended or unanticipated consequences of adopting a particular approach (in this case, a structural shift from existing boundaries that are political to those that are bioregional).

The difficulties of persuading the citizenry of the need for changes that would bring about an ecological consciousness are manifold. Foremost, perhaps, is the fact that there are powerful myths and metaphors that guide our thoughts and actions and that compete with the notion of ecological consciousness. In Western, industrialized societies, there are deeply held notions about the sanctity of the individual self, the importance of economic freedom, and the power of technology to enhance our lives. While I would agree that "the principles of paradigms...only allow [people] to see what they are willing to see" (Hynes 1994, 11), I take issue with the notion that changing the paradigm "can happen within one second, within one single person who suddenly sees things in a wholly different way" (Meadows 1972, 11).

On the one hand, profound revelations or insights can occur in a matter of moments when a new relationship or pattern is discerned. One example might be a powerful single image, such as Earth seen from outer space, which led many to see our planet as a beautiful but fragile biosphere. Another might involve a thoroughly rational explanation, such as the notion that cooperation among species is as prevalent as competition. However, we live in deeply contextualized perceptual worlds where culture, language, and thought exercise a restraining power on insights that at first glimpse might seem to question dominant paradigms. Typically, we return to familiar ways of thought and action while holding in abeyance those notions that seem not to fit our ordered perceptual worlds. Perhaps most North Americans exhibit this cognitive ambiguity. They have heard and seen enough to wonder about the fragility of our environment, but they do not yet see that our underlying cultural values may be to blame. On the other hand, those who cognitively see the destructiveness of our present way of life also face the terrible realization of a dearth of career opportunities that will actualize their ecological sensibilities.

Problems with Adoption of Bioregionalism

Beyond the difficulties inherent in adopting any notions that are radically opposite to the dominant cultural paradigms, there are specific problems with the notion of bioregionalism—at least as envisioned here as the redrawing of boundaries along lines coterminous with the bioregions themselves. I shall briefly indicate these:

• Despite the fact that present county and state lines make little or no sense in terms of their relationship to local ecosystems, larger bioregions, or even "ecoregions" (the Sierra Club has begun an initiative that focuses on the 21 "ecoregions" of North America), there is something to be said for the known political and economic realities that such boundaries have helped demarcate over time. I am well aware that a primary mission is to change political and economic realities, but (as will become clearer later), I do not believe that re-drawing the map is worth the effort in a strategic sense.

• Even if the United States could decide on what the bioregions of the country should be (and this would be no small task, given the competition for the natural resources that presently exist), it is not clear that shifting to bioregional boundaries would necessarily reduce present tensions. The fact that a particular watershed would now be placed within one bioregion rather than scattered over several states does not mean that those states lose interest in the water resource.

• An even larger concern for most U.S. citizens would be the whole issue of national boundaries. If one were true to the whole idea of bioregionalism, one would have to concede that the borders with Mexico and Canada are also artifacts of history rather than geography. Where would it end? Throwing open such borders (actually doing away with them) would present immense political, legal, social, and economic problems. Furthermore, though self and community identity as coded by state residence has eroded in most places over time, there is still an intense identification with one's country, and this is true for the citizenry of the United States, Mexico, and Canada.

• A major complicating factor with any proposed systemic change that attempts to address the root causes of environmental degradation is the presence and economic dominance of national and multinational corporations. Perhaps it is unfair to place this concern under the banner of problems specific to bioregionalism, but as a bastion of decentralization, bioregionalism calls attention to the disparity between small-scale, local operations in tune with the environment and the realities of worldwide trade, information superhighways that mitigate against strong connection to a place, and the dominance of multinationals in the economic marketplace.

• Though it may also be unfair to fault bioregionalism for not articulating a clear vision as to how greater social justice would ensue with such an approach, this element does not appear to be in the foreground. Accordingly, the idea does not have the appeal that could transform the rising unrest among significant sectors of this society into a unified assault on other problems, which also include poverty, crime, and health issues, all of which need to be part of systemic change.

Conclusions

Despite my seeming comments to the contrary, I am enamored of the bioregional vision—but more for its educative, visionary function than as an actual policy for environmentalists to put all their weight behind. It does represent a rational way of integrating the human sphere within the larger biotic community, and I do believe that "mapping the bioregion" (using whatever an individual or community deems to be its bioregion) is an excellent way for people to see their communities in relation to their natural surroundings. Yet bioregionalism presents too many "red flags," and with precious time for change slipping away, it would be a strategic mistake for environmentalists to invest in this approach. Perhaps more relevant may be what Mark Dowie refers to as the human rights approach to environmental advocacy, which connects the rights of individuals to life and the pursuit of happiness with freedom from the "social and environmental devastation wrought by polluting industries" (Dowie 1995, 244). Here we have the kind of language that seems more resonant to our ears. In fact, if we took the notion of connecting human rights with ecological health and used the language and spirit of community empowerment and decentralization of decision making (both hallmarks of bioregionalism), we might yet forge an effective and comprehensive alternative to what presently exists.

References

Aberley, D., ed. 1993. *Boundaries of Home.* Philadelphia: New Societies Publishers.

Doughty, R. 1990. "Nature Writing and Environmental Experience." In L. Zonn, ed. *Place Images in Media.* Savage, MD: Rowman & Littlefield.

Dowie, M. 1995. *Losing Ground.* Cambridge, MA: MIT Press.

Elder, J. 1994. "The Big Picture." In *Sierra* (March-April), 53.

Hynes, M.E. 1994. "Walking in a World of Wounds: The Work of Donella Meadows." In *CenterPiece* (Fall), 11.

Meadows, D. 1972. "Limits to Growth." In *CenterPiece* (Fall), 1994, 11.

Chapter 10

Noise Pollution: A Growing World Problem

Arline L. Bronzaft
Professor Emerita, Lehman College
The City University of New York
and
Council on the Environment, New York City

New York City has been described at its worst as dirty, crowded, and noisy, but according to a September 1996 poll (Allen 1996), noise is ranked as that city's most disturbing "quality of life" issue. However, the noise that permeates the environment of this metropolis and penetrates into the living quarters of its citizens is not a unique phenomenon. Cairo's modes of transportation contribute significantly to that city's high noise level; Athens's automobile din is similarly disconcerting; and one out of three people in the United Kingdom stated that "environmental noise spoiled their home life" (Grimwood 1993).

Two thirds of Germany's residents complain about excessive noise, especially from traffic (*The Week in Germany* 1996). Hartmut Ising's January 30, 1996 radio program discussing Germany's noise problem was aptly named *Krankmacher Lar* ("Sickening Noise"). Highlighting the growing noise problems in Britain, Canada, and the United States, the Canadian Broadcasting Company's *Witness* series opened its 1996 TV season with a program entitled *Sound and Fury* (Valpy 1996).

Noise is not simply a matter of excessive street sounds or increased air traffic; it is the result of urban living that brings ever-larger numbers of people into close proximity. Add to this the fact that individuals have grown less respectful of other people's rights to "peace and quiet," and a fuller picture of the noise pollution problem emerges. However, as noted by Zaner (1991), sounds that have a known adverse effect on human beings have been present as environmental pollutants for thousands of years. Zaner describes accounts of noisy delivery wagons on the streets of ancient Rome,

Old Testament stories of loud music, and old poems and stories of loud and annoying animals.

After the Industrial Revolution, noise became associated with large urban areas. In recent years, noise has invaded even the quietest of communities. Noisy leaf blowers and lawn mowers have made suburban living less desirable. Small-town residents find themselves thrust into controversies over the siting of new highways adjacent to their homes.

Many city residents have traditionally felt that noise is simply part of life in the "big city." City dwellers learned to live with the noises that surrounded them—traffic, overhead jets, screeching trains, and loud neighbors. Those who desired a quieter environment were encouraged to move away from the bustling cities and commute to their jobs daily from the suburbs. In time, however, even these once quiet suburban towns became engulfed in offensive sounds. Cities grew noisier as passing trains became louder, the numbers of overhead jets increased, and automobile traffic began to clog roadways day and night. Apartment buildings got taller, and greater numbers of people were navigating narrow streets, shopping in neighborhood stores, and competing over limited parking areas.

Were these urban dwellers to accept the noises surrounding them as the price for living in these areas? Or had noise reached levels that were distressing to even the most tolerant urbanites? Surveys had found that noises were an irritant (Bronzaft & Madell 1991; Schultz 1978). Were these noises also affecting the wellbeing of city dwellers? Was noise a mental and physical health hazard? To assess whether noise was indeed a health hazard, scientists would have to conduct studies of the type done earlier to assess the impact of cigarette smoking on the body.

Research studies suggest that noise pollution can be a sufficiently serious health hazard to warrant government legislation. The President of the Health Council of the Netherlands has called the "abatement of noise annoyance, of noise-induced hearing loss, and of other effects of noise on health…an important part of public health policy" (Health Council of the Netherlands 1994).

In many countries, public health policies are based on data collected when noise exposure was less than it is today. Despite the fact that recent studies on the effects of noise on health require further validation, existing environmental and health policies would benefit greatly from the knowledge generated by this research. Before elaborating further on public policy, let us look to the data we presently have concerning the effects of noise on mental and physical health.

Defining Sound and Noise

We must differentiate between sound and noise. Noise can be defined as unwanted sound. Noise is also uncontrollable and unpredictable. Sound travels through the air in waves and has two major characteristics: the speed at which the waves vibrate and the intensity of each vibration. For the most part, intensity accounts for the loudness of sound, though frequency contributes to some extent, with higher sounds perceived as louder. Sound is measured on a decibel scale, which is actually a physical measurement of sound pressure. To allow for the effect of frequency on loudness, a modified form of the decibel scale, the A scale (abbreviated as dBA), is used to measure loudness as people hear it. Additionally, the "loudness" of different sounds has been determined by asking individuals to listen to them and give their judgments as to how loud they are. When loudness is measured as humans hear sounds, an increase in 10 decibels represents a twofold increase in sound, 20 decibels, a fourfold increase, and so on. The decibel scale is logarithmic, not linear.

Some common sounds have been measured as follows: leaves rustling 10 dBA, human conversation 50 dBA, vacuum cleaners around 90 dBA, subway trains 100 dBA, and loud music over 120 dBA. Sounds that are over 60 dBA may be judged as intrusive, sounds over 80 dBA are deemed to be annoying, and sounds over 100 dBA can be extremely disturbing. However, when sound acts as noise, it need not be loud to be bothersome. The drip of a faucet, the low bark of a neighbor's dog as one is reading, and the low but distracting TV set next door as one is falling asleep are sounds defined as noise. All sounds are reacted to psychologically, as well as physically. With noises, the psychological components contribute significantly to their impact on individuals.

The Effect of Loud Sounds on Hearing

What is the effect of loud sounds on hearing? Sufficient data support the statement that loud sounds can damage hair cells in the ears, the receptors for sound. A single loud sound, such as an explosive, can cause permanent damage. However, hearing loss is more frequently the result of continuous exposures to loud sounds. Studies support the relationship between long-term exposure to loud sounds, generally over 85 dBA, and hearing loss (Kryter 1994; Fay 1991). A 1990 National Institutes of Health report notes that approximately 28 million Americans suffer from some hearing loss, and approximately 10 million of these impairments have been partially attributed to damage from loud sounds. This report warns that more than 20 million Americans are regularly exposed to noise levels that could lead to hearing loss.

The American government has established guidelines through the Federal Safety and Health Act to protect individuals working in noisy industrial settings, but not all workers wear the ear plugs and earmuff-type shields recommended. Loud sounds are no longer restricted to the industrial plant, and more and more people are exposed to extremely high decibel levels in the recreational setting—at rock concerts, discos, and video arcades. Youngsters are also listening to their headsets at very high levels (Madell 1988). Dr. Marjorie Downs (Newcomer 1988), a noise expert, concluded that "baby boomers whose teenage summers revolved around rock concerts aren't hearing as well as they used to, and their children are repeating their mistakes." More distressing are companies whose advertisements promote loud sounds. Nintendo urges "playing it loud." A Toronto brewery airs a screaming advertisement proudly stating that it is "currently in violation of every noise bylaw ever written by City Hall."

Sound as Noise and its Effects on the Body

Whereas damage to hair cells is a direct result of loud sounds, sounds of any intensity, identified as noise, can affect us indirectly. The body reacts to noise as it does to other disturbing stimuli—with increases in blood pressure, excessive secretions of hormones, and changes in heart rhythms. What happens if noise continues unabated? There is a growing body of literature on the nonauditory effects of noise, largely examining the impact of noise from highways, elevated trains, and airports. Such studies have looked at cardiovascular, circulatory, and gastrointestinal ailments resulting from noise (Evans, et al., 1995; Kryter 1985, 1994; Passchier-Vermeer 1993; Fay 1991; Tempest 1985).

A recent wellness questionnaire (Bronzaft et al. 1996) containing questions on noise was distributed to two communities, one within the path of a New Jersey airport and the other in a quieter community. The data indicate that significantly more people living in the flight paths were disturbed by noise, and those who reported themselves bothered by air-flight noise were also more likely to perceive themselves as being in poorer health. In contrast, no relationship between being bothered by noise and the perceived state of health was found in the quieter area. While the studies cited require further validation, as Susan Staples (1996) points out, the evidence suggests that noise adversely affects health.

The Effect of Noise on Sleep and Good Mental Health

In his review chapter, Pollak (1991) cannot conclude that noise-induced sleep loss harms health; he does find that sleep loss impairs task performance the next day. Generalizing from this finding, we might hypothesize that inability to sleep well might make it more difficult to carry out tasks the next day. Sleepiness may also make people less attentive to warning signals in the environment—horns honking or sirens blaring.

The Bronzaft et al. study (1996) cited above also found that residents living in the path of aircraft traffic patterns who reported being bothered by noise also stated that they had more sleep difficulties. There was no relationship between sleep difficulties and being disturbed by noises in the community not subjected to overhead jets.

Abey-Wickrama et al. (1969) and Herridge & Chir (1972) reported higher admissions to mental hospitals among people living near noisy airports, but the methodology of these studies has been questioned, so that additional research is called for along this line. However, there is no denying reports of complaints from individuals living near airports who express their frustration by saying the noise is "driving them mad" (Onishi 1996). Good mental health can also be measured in terms of how much aggression people express in describing their feelings toward the sources of the noise, namely the airport managers or noisy neighbors. Laboratory studies suggest that noise increases aggressive behaviors, but even more pertinent are the numerous stories in the media indicating fights between neighbors because of noise. In 1995, London newspapers highlighted cases of at least 20 people who died in disputes with neighbors over noise. Additionally, there were reports of people committing suicide because of their noisy neighbors.

Noise Affects Child Development and Learning

Bronzaft (1991) reviewed studies examining the effects of noise from nearby highways, elevated train tracks, and airports on children's learning and on their reading skills in particular. Lower reading scores were found for children attending schools in the path of noisy transportation. Evans & Lepore (1993) reported that children attending schools near airports often give up more readily on tasks. Such behavior has been identified as learned helplessness, the tendency to quit trying when you believe you can't control what is happening to you.

Children can be harmed by noise within their homes. Wachs & Gruen (1982) found pre-school children's language and cognitive development were impeded in very noisy homes. They also found that noisy home environments did not foster good communication between parent and child. By contrast, older adults (who had been outstanding students) recalled that their childhood homes had provided them with the quiet environment that was favorable to reading, studying, and learning. Their parents had created an overall quiet atmosphere for them, and undoubtedly, the quieter home life they described contributed significantly to their good study habits, as well as to their overall healthy development (Bronzaft 1996).

Noise Intrudes on Oases of Peace and Quiet

When the Park Service was created in the United States in 1916, it was charged with protecting our environmental legacy so that Americans and foreigners alike could gain a sense of serenity and pleasure as they contemplate the natural wonder of our national parks. But few today would call the Grand Canyon peaceful and quiet as over 10,000 planes and helicopters fly over the canyon in the summer months (Jaroff 1995). Jet skis along the lakes of American cities are depriving visitors of the enjoyment afforded by these waterways.

Enough Data to Initiate Noise Abatement-But There are Problems!

Although further research demonstrating the hazardous effects of noise pollution must be undertaken, there appears to be enough evidence to support intense and unremitting efforts to abate environmental noise. Research findings that noises interfere with learning should lead to requirements that schools not be placed near busy highways, near elevated train tracks, or in noisy industrial settings. At the least, provisions should be made to shield classrooms from exterior noises through the use of noise-abatement materials. For example, when it was found that noise from nearby elevated tracks lowered the reading levels in a school in upper Manhattan in New York City (Bronzaft & McCarthy 1975) rubber pads placed on the tracks and acoustical ceilings put into the ceilings of the rooms facing the noisy trains lowered the decibel levels in the affected classrooms and raised children's reading scores (Bronzaft 1981).

Aircraft noise affects millions of people in the United States, but the government response to the plight of these people might be judged to be negligible, especially if one were to query the over 200 groups that have been formed to combat aircraft noise (ARCO 1995, personal communication). The U. S. Congress acknowledged an existing noise problem in its passage of the Aviation Safety and Noise Abatement Act in 1979, which established a program by which airports could apply for federal funds to mitigate noise. Such measures included soundproofing residences and schools near airports. Unfortunately, New York City with three major American airports, including an airport with the largest population living within the critical noise contour, decided not to participate in this program (Stenzel 1996).

It should be noted that a group of Japanese citizens living near an American air base in the Tokyo suburbs is suing the U.S. government (Kristof 1996). The plaintiffs are demanding a ban on flights between 9 p.m. and 6 a.m. and $30 million in damages. Leaving aside the political implications of such a lawsuit, it should hold special interest for the millions of Americans living near airports in the United States.

Noise Abatement: A New Goal for Environmental Activists

As chairman of the Noise Committee of the Council on the Environment of New York City, I have worked with community groups formed to curtail noise. Some groups are more concerned about neighbor noise, others about aircraft noise, and still others about train noise. They share one thing: a belief that the government is not doing all it can to reduce noise. These citizens have educated themselves on the dangers of noise and are attempting to educate the larger population.

In New York City, the Big Screechers organization was effective in lessening the din of passing elevated trains. ARCO in Chicago is working to quiet overhead aircraft. One suggestion coming from groups against air-craft noise deals with imposing a limit on the number of short-distance flights. For example, does New York need a plane to Detroit, Washington, and Toronto every hour on the hour? The Campaign for Peace and Quiet in the Bronx is battling noise by appealing to churchgoers especially. The leader of this campaign can be heard on Sundays asking fellow parishioners to lower the decibel level of a borough he claims has been under acoustic siege (Dallas 1995).

Two Canadian groups, the Citizens' Coalition Against Noise in Toronto and the Right to Quiet Soundscape Awareness in Vancouver, are confront-ing community noises. The Noise Network in London is causing the press to pay greater attention to noise pollution.

Informed public officials can be instrumental in instructing govern-ment agencies to provide the requisite funding for research into the health impacts of noise. The present evidence, suggesting the relationship between adverse health effects and noise, may not be sufficient to prod the majority of legislators to pass stronger anti-noise legislation, but the data are con-vincing enough for knowledgeable officeholders to seek research dollars. If additional research should support the already suggested relationship be-tween noise and health, then we could, hopefully, expect the passage of laws in keeping with the data.

However, scientists can also play an important role in enlightening of-ficeholders on the need for further noise research. Susan Staples (1996) calls for psychologists to increase their involvement in environmental noise research. Those psychologists who have already undertaken some worth-while noise experiments could take the initiative and engage legislators in meaningful dialogues on their findings as well as discourse on how psycho-logical research could benefit public policies relating to the noise issue.

Conclusions

Not very long ago, when concerned individuals attempted to persuade leg-islators that cigarette smoking was harmful to health, tobacco growers were even more persuasive in their arguments advocating the safety of their prod-uct. Without the required research to negate these arguments, the tobacco industry continued to market dangerous products. Let us learn from our experiences with tobacco and work to abate the pervasive and perilous pol-lutant of noise. Governments worldwide must coordinate their efforts to establish environmental and health policies that mirror recent scientific and technological knowledge in the field and, at the same time, fund additional research. Government efforts to curb noise pollution should advance the wellbeing of all inhabitants of the planet.

References

Abey-Wickrama, M.F. a' Brook, F.W.G. Gattoni, & C.F. Herridge. 1969. "Mental Hospital Admissions and Aircraft Noise." *Lancet 2,* 1275-77.

Allen, M. O. 1996. "Gripes on Noise are Burning up New Cop Hotline." *The New York Daily News* (September 30).

Bronzaft, A.L. 1996. *Top of the Class.* Norwood, NJ: Ablex.

_____. 1991. "The Effects of Noise on Learning, Cognitive Development and Social Behavior." In Fay (below), 87-92.

_____. 1981. "The Effect of a Noise Abatement Program on Reading Ability." *Journal of Environmental Psychology 1,* 215-22.

Bronzaft, A.L. & J.R. Madell. 1991. Community Response and Attitudes toward Noise. In Fay (below).

Bronzaft, A.L. & D.P. McCarthy. 1975. "The Effect of Elevated Train Noise on Reading Ability." *Environment and Behavior 7,* 517-28.

Bronzaft, A.L., J.A. O'Connor, K. Ahern, R. McGinn, B. Savino. "Aircraft Noise: A Potential Health Hazard." (Submitted for publication).

Dallas, J. 1995. "No more Jerichos!" *Hearing Rehabilitation Ouarterly 20,* 9-11.

Evans, G.W., S. Hygge & M. Bullinger. 1995. "Chronic Noise and Psychological Stress." *Psychological Science 66,* 333-38.

Evans, G.W. & S.J. Lepore 1993. "Nonauditory Effects of Noise on Children: A Critical Review. *Children's Environments 10,* 31-51.

Fay, T.H. 1991. *Noise and Health.* New York: The New York Academy of Medicine.

Grimwood, C. J. 1993. "Effects of Environmental Noise on People at Home." Garston, Watford, UK. Building Research Establishment.

Health Council of the Netherlands. 1994. *Noise and Health.* The Hague: Health Council of the Netherlands.

Herridge, C.F. & B. Chir.1972. "Aircraft Noise and Mental Hospital Admission." *Sound 6*, 32-36.

Jaroff, L. 1995. "Crunch Time at the Canyon." *Time* (July 3), 40-41.

Kristof, N. D. 1996. "A Lawsuit in Japan Seeks to Ban Noisy Flights at U.S. Base." *The New York Times* (April 11).

Kryter, K.D. 1994. *The Handbook of Hearing and the Effects of Noise.* San Diego: Academic Press

_____. 1985. *The Effects of Noise on Man.* Orlando, FL: Academic Press.

Madell, J.R. 1988. "A Report on Noise." *Hearing Rehabilitation Quarterly 11*, 4-13.

Newcomer, K. 1988. *Rocky Mountain News* (February 19) Denver, CO, 7.

Onishi, N. 1996. "Battle lines in a War on Noise Lie Behind the Runway." *The New York Times.* (August 15), B1, B5.

Passchier-Vermeer, W. 1993. *Noise and Health.* The Hague: Health Council of the Netherlands.

Pollak, C.P. 1991. "The Effects of Noise on Sleep." In Fay (above), 41-60.

Schultz, T. J. 1978. "Synthesis of Social Surveys on Noise Annoyance." *The Journal of the Acoustical Society of America 64*, 377-405.

Staples, S. 1996. "Human Response to Environmental Noise: Psychological Research and Public Policy." *American Psychologist 51*, 143-50

Stenzel, J. 1996. *Flying Off Course: Environmental Impacts of America's Airports.* New York: Natural Resource Defense Council.

Tempest. W. 1985. *The Noise Handbook.* Orlando, FL: Academic Press.

The Week in Germany. 1996. "Environmental Agency Cites Noise as Germany's Least Recognized Environmental Threat" (February 9). New York: German Information Center, 6.

Valpy, M. 1996. "Leaf-blowers and the GDP." *The Globe and Mail,* Toronto, Canada (September 27), A21.

Wachs, T. & G.E. Gruen. 1982. *Early Experience and Human Development.* New York: Plenum.

Zaner, A. 1991. "Definition and Sources of Noise." In Fay (above), 1-14.

Chapter 11

Military Activities: Their Impact on the Environment

Jack Ahart
Captain, U.S. Navy (Ret.)

The U. S. military prides itself on being a professional force. Its primary mission today remains much as it was in the past: to protect the vital interest of the United States against all enemies. By means of an all-volunteer force, the four services of the Department of Defense (DOD) combine with their Department of Transportation counterpart, the Coast Guard, to ensure that no foreign aggressor could ever successfully attack the United States. Members of this dedicated ensemble are undoubtedly the best educated and most technically advanced military force America has ever produced, at least when it comes to "force projection" and "conflict management" (today's buzzwords for combat).

The instruments of today's military might are the likes of the deadly M1-A2 Abrams tank, the extremely accurate F-117 Stealth fighter, the ominous Trident submarine, impressive amphibious forces, and (being a naval flyer myself), of course, the carrier battle groups. Additionally, the military's reliance on an all-volunteer force guarantees the United States a wealth of intelligent and dedicated women and men to employ these tools of destruction if world events should dictate their use. The advantage of America's military might in Operation Desert Storm is a testament to just how effective today's military machine can be in accomplishing its primary function of protecting the country's vital interests. Even so, a definite change in the military's mission is occurring.

The end of the Cold War and decreasing world tensions are allowing the United States to reduce the size of its military forces and infrastructure. By the end of 1992, 122 bases in the United States, from all four services, were programmed for closure, and even more closings are proposed for the future. In addition, reduced tension in Europe and other countries is now permitting forward-deployed military forces to return to the United States. Military "downsizing" will result in the United States' closing 492 overseas bases while reaping huge "peace dividends" in the process (Satchell 1992, 28). No doubt this is welcome news for the struggling American (and world) economy. In the United States, Americans anticipate reclaiming billions of dollars as the assets of closed bases are returned to state and local municipalities. The DOD estimates that $3.1 billion will be saved annually by these base closures (Palmer 1993, 770).

Overseas, the U.S. government uses a more complex formula to compute anticipated "peace dividends" as our troops withdraw. The military first totals the value of "improvements" the United States made in the respective base, subtracts the cost of environmental damage attributed to the American presence, and then calculates the difference. Not surprisingly, the United States winds up with a positive balance. Utilizing such oversimplified accounting practices leads the uneducated to the conclusion that all the United States must do is reduce the size of its military and count the money saved.

Events of the past few years, however, indicate that perhaps the "peace dividend" will never materialize. What is worse, there may be a Cold War legacy so dangerous that it will take generations and enormous financial commitments to rectify.

A Legacy of Environmental Devastation

In many respects the military is not unlike a large business such as General Motors or IBM. The military certainly has a culture of its own. It also has a stratified personnel pool and a definite hierarchy that provides a means of advancement and development for its members. It also requires sophisticated leadership and management skills found in successful corporations. Just as a civilian corporation produces a product, so does the military. Unlike General Motors, however, the products of the U.S. armed forces aren't always consumable or measurable. An example of an intangible product would be the freedom of navigation of the high seas. This may sound nebulous, but it is critical to the economic and political prosperity of the United States and is but one of the products (missions) the U.S. Navy provides. Each service produces similar important yet intangible products. Also, like its civilian counterparts, the military produces waste byproducts (including toxic wastes) during the accomplishment of its missions.

Environmental Accountability Comes to the Military

How is the military held accountable for the "proper" disposal of these waste byproducts? In recent years, two important actions taken by the federal government changed the way the military thought about its environmental responsibilities.

In 1986, an amendment to the 1980 Comprehensive Environmental Response, Compensation, and Liability Act (CERCLA) made previously exempt federal facilities environmentally accountable. Later, in October 1992, the Federal Facilities Compliance Act (FFCA) changed the way in which the military does business. The FFCA makes the Pentagon liable for any future legal challenges that flow from the toxic products it might leave behind. These laws forced significant changes in how the military would conduct its business.

No longer shielded, the military received notice from Congress that its commanders would be held responsible for future violations as well as for mismanaged cleanup efforts at pre-existing sites. To drive this point home, a number of individual commanders and subordinates have been prosecuted under new federal, state, and local laws. In the case of *United States* v. *Carr*, an Army civilian at Fort Drum, New York, received a one-year suspended sentence, a year's unpaid suspension, and a demotion for instructing his subordinates to dump and bury cans of waste paint. In *United States* vs. *Dee*, three Army civilian scientists at the Aberdeen Proving Grounds were convicted of failing to properly dispose of hazardous waste generated by their laboratory. Each received three years' suspended probation, as well as 1,000 hours of community service. Another indication that a casual approach to environmental stewardship is unacceptable to the military is reflected by the ten-month prison sentence John Hoyt Curtis received for spilling more than 500,000 gallons of jet fuel at Naval Air Station Adak, Alaska. For his complacency, Curtis received a jail sentence, as well as a federal criminal record reflecting one felony and two misdemeanor counts of violating the federal Clean Water Act.

Environmental Ethics and the Military

How should the military begin to heal the plethora of environmental wounds that the United States created during the Cold War? Perhaps a wise starting point would be to review the 1972 Declaration of Principles drawn up by the United Nations Conference on the Environment. This basic statement points the environmental finger of responsibility at the United States, as well as at every other nation. These 26 principles start with the simple statement: "Man has the fundamental right to freedom, equality, and adequate conditions of life, in an environment of a quality which permits a life of dignity and wellbeing, and bears a solemn responsibility to protect and improve the environment for future generations...."

Certainly these environmental rights would include, as a minimum, the following issues: clean air and water; healthy living space; essentials of life (food, clothing, and shelter); job opportunities; a healthy occupational setting; world resource sharing; and knowledge of environmental threats. While not all of these items are affected by the environmental issues for which the military is responsible, the tone and flavor of such environmental rights are clear. People on Earth have the right to live in an environmentally friendly world.

As a world leader and a charter member of the United Nations, the United States has a moral responsibility to correct its legacy of environmental mismanagement. While the United States is only one factor in the global environmental equation, it must assume an environmental leadership role, much as it would in any peacekeeping, economic, or diplomatic role. It is the only ethical answer.

Ironically, some individuals would argue from a utilitarian ethical position that, during the Cold War, the military was in fact "providing the greatest amount of good for the greatest amount of people" (Frankena 1973, 34) by providing a competent force that prevented nuclear conflict. Proponents of this position hasten to elaborate that without the efforts and preparations of the armies of the free world, a more destructive fate would have threatened the planet. They also argue that the environmental challenges facing us are unfortunate but necessary byproducts of the U.S. military's successful efforts to prevent a greater evil. Invariably, in the next breath these individuals attempt to allay ecological concerns by theorizing that advancing technology will provide deliverance from environmental nightmares.

Opponents of this viewpoint believe this to be a convoluted and flawed argument. For them, simply, there can be no ethical argument for the environmental devastation that has occurred and is still occurring at the hands of the DOD.

What Can Be Done?

As horrendous as the government's environmental track record is, the women and men of the U.S. military have several tools at their disposal to help remedy the situation. The first of these tools is the available core of educated and disciplined personnel inherent among the military. These individuals are professionals who can translate the energies of armed conflict into the mundane but perhaps more important environmental remediation process. While environmental activism has yet to become the military's new mission, a transformation is now underway as newly evolving roles for today's military include disaster relief (e.g., Hurricane Andrew relief efforts) and educational assistance to inner-city schools, as well as international peacekeeping and peacemaking efforts. These activities are nontraditional military roles and are envisioned as adjunct roles or missions. They are not designed, or proposed, as replacements for the traditional primary military role of protecting the vital interests of the United States against any aggressor. The question to be considered is: Are environmental interests also vital interests?

Logic would suggest that an additional nontraditional role focusing on environmental issues could, and should, be developed. This environmental mission would permit the military to assist in protecting and regenerating the environment. More specifically, the Air Force would patrol and report on remote areas of the world, forest fires, and tracking of endangered species. Navy personnel and assets would play an instrumental role in monitoring pollution of the oceans and overfishing in international waters (Lewis 1992, 47). Realistically, the military is not ideally suited to such radically different roles. However, with training and leadership, the military could focus its activities and enormous assets on nontraditional roles, including environmental monitoring. No one is suggesting that Abrams tanks or M-16 rifles have a part in reversing the ecological damage of the past. What is being proposed is a realistic utilization of research facilities, aircraft, ships, and personnel in beneficial environmental activities. To do so, however, will require major changes in the current military culture and mindset.

Adjustment of the military mindset has already begun. Current military jargon now includes the environmental buzzword "stewardship" as America's armed forces are jumping into the environmental business with both feet. Commanders are now painfully aware that Congress intends to hold them personally accountable for the ecological failings of their organizations. Considerable energies and resources are now being marshaled within each of the services in a belated effort to tackle Herculean environmental challenges. Of these, restoration of contaminated facilities is the greatest challenge facing the DOD. In order to mitigate existing environmental damage, the DOD established the Installation Restoration Program (IRP) and the Formerly Used Defense Sites (FUDS) Program. Virtually all cleanup activities addressed by these programs will be conducted through civilian contracting and, given congressional funding constraints, performed on a "worst first" priority basis.

Leadership and Military Environmental Activism

Active involvement by a commander in the mission of any organization is critical to success. Without her or his sincere interest in an organization's activities, little can be accomplished. Such is the organizational culture of the military. This same culture, combined with aggressive leadership, can also work to galvanize units as they tackle the most demanding tasks (or missions). So will it be when conducting the military's environmental business in the future.

Unrelenting reports of environmental tragedies caused by irresponsible military commanders continue to flood the media. There is no denying that the military has caused significant environmental damage, often through ignorance but frequently through inertia. For these transgressions, the military will pay dearly in the form of eroded public trust and severe drains on decreasing fiscal budgets as more and more remediation comes out of day-to-day operating costs. A contaminated tennis court-sized area costs $17 million to clean up! Therefore, installation commanders can no longer afford to be complacent about the environment they oversee. If they are, they will face the harsh realities of civil litigation with little, if any, military legal assistance. The military as an organization and today's military commanders are now expected to be responsible environmental partners who no longer hide behind the shield of the Cold War. Communities will no longer stand by and allow their water, soil, and air to be polluted by irresponsible members of these communities.

Conclusions

Commanders must be the focal point for implementing sound environmental practices by their units. They must also provide guidance in the form of leadership, short-range and long-range plans (Commander's Environmental Guide), and an environmental vision to focus the energies of their educated, dedicated, and energetic personnel. Environmental inactivity is not and will not be tolerated. Much damage has already been done, and it will take a concerted effort by the military to rectify the situation. The cleanup process will be expensive and arduous, but it must continue at an ever-increasing pace.

The stark reality is that the military must still protect the vital interests of the United States against military aggressors, but it must also make war on its environmental problems. Environmental stewardship must and shall be the new mission of the military commander in the 21st century!

References

Garcia, Elroy. 1993. "Restoring the Earth," *Soldiers 48*, 4 (April): 18-19.

Frankena, William K. 1973. *Ethics*, 2d ed. Englewood Cliffs, NJ: Prentice-Hall.

Lewis, Damien. 1992. "Soldiers on Eco-Patrol," Gemini News Service of London in *World Press Review 39*, 11 (November).

Palmer, Elizabeth A. 1993. "Pollution Clogs the Transfer of Land to Civilian Uses," *Congressional Quarterly 51* (27 March).

Satchell, Michael. 1992. "The Mess We've Left Behind," *U.S. News and World Report 113*: 21 (30 November).

Chapter 12

A Business Approach to Environmental Education in Canada

Glenn D. Ruhl
The Environmental Services Association of Alberta
Alberta, Canada

The links among business, environmental policy, and employment are beginning to manifest themselves in North America. Environmental employment, according to the Canadian Council for Human Resources in the Environment Industry (CCHREI), involves the business activities associated with environmental protection, the conservation and preservation of natural resources, environmental education, environmental communications, and environmental research (CCHREI 1994, 6). In the business community (known as the environment industry), EE is associated with those educational activities that promote the growth and development of one's business.

A newcomer to the environment industry soon realizes the industry's focus is distinctly *business* rather than *green*. Not surprisingly, until recent times, industry rarely exhibited an environmental conscience. In the past, industry was characterized by a stoicism that viewed the Earth as a machine to be used and discarded at will. Industry considered that all was made for the sake of humankind, or, as Descartes propounded, "we are the masters and possessors of Nature" *(Discourse on Method* Part VI). Today, however, a cooperative and balanced understanding is emerging. Human "use" of the planet's resources is now constrained by larger ethical considerations. What has emerged from this constraint is an environmental awareness that has become an exciting opportunity for new business ventures.

The Canadian environment industry is composed of small to medium-sized businesses.[1] In general, these businesses, because of their size, benefit greatly from association membership, which enables small companies to form the "critical mass" required to compete at a significant level with large corporations. The Environmental Services Association of Alberta (ESAA) is a not-for-profit business association of over 370 companies providing environmental services.[2] Referred to by Industry Canada as a "model association for the 1990s," the ESAA is the strongest environmental industry association in Canada. Its member companies employ over 10,000 people and generate over 1 billion dollars in annual sales in a province of 2.6 million people. The ESAA provides its members with a wide range of programs and services through four main strategic thrusts: education, technology, information services, and marketing. In all cases, the ESAA adheres to a mission dedicated to building a strong environmental-services industry through leadership in these strategic areas.

The environment industry is a growing force in the Canadian business community. In Alberta, the industry's strength is reflected in the ESAA's educational programming initiatives which are a direct response to the needs identified by industry stakeholders both provincially and nationally. The ESAA's work in the area of human resources has grown over the past five years from workshops and seminars to course offerings, training programs, and support for certification and accreditation. Recently, these initiatives have included training projects and/or professional development for (1) unemployed science and technical professionals, (2) 18- to 24-year-olds for careers in the environment industry, (3) marketing professionals, and (4) various industry practitioners. As a result, Alberta's environment industry has developed mutually beneficial working relationships with universities, colleges, and both federal and provincial governments. This synergistic effort far exceeds the influence that might be achieved by any one component working alone. A business approach to EE enables industry, government, and the public to clearly see the advantages of working together in terms of what makes sound business sense, as well as, what makes sense environmentally.

Background

Rachel Carson admonished us some 30 years ago in *Silent Spring* (see Corcoran, this volume) that if we do not learn to live in harmony with the natural world, we will destroy it and ourselves. It is the public's demand for a safe, clean environment that created the need on the part of corporations to re-evaluate and re-order their priorities with respect to environmental issues. The environment industry arose out of this need and has, as its primary task, a duty to carry out these priorities. It is ironic, perhaps, that the environment industry continues to wince somewhat when the term "green" is used. The environment industry is the result of the same "environmentalist" philosophy that spawned the "green" movement, and the legislation and regulatory affairs that drive the environment industry are a direct spinoff of the continuing role played by environmentalist lobby groups and their affiliates.

The Canadian environment industry is a growth industry within a national economy facing serious challenges. Although the Canadian government subsidizes every college and university student at a rate of almost $16,000 per annum, 18 percent of Canadian youth do not complete high school. Over 90 percent of Canadian dropouts have not completed grade 10, a minimum of 24 percent of Canadians aged 18 and over are functionally illiterate, and many are unable to do simple arithmetic (Royal Bank 1995). The impact of these alarming statistics is further emphasized when one considers that the unemployment rate for those Canadians with a university education is 5.1 percent, compared with 14.1 percent for high school graduates (*Calgary Herald* 1995, A5). It is now obvious that even entry-level work will require increasingly stringent educational requirements. Still, 60 percent of young Canadians enter the job market with no post-secondary or vocational training.

A growth industry demands personnel with essential industry-specific skills, training, and knowledge in order to maintain a competitive position within the business community. In 1992, an examination of the environment industry's critical human resource needs identified the issues of vital importance to the Canadian environment industry. The industry noted a lack of management skills, particularly among small, high-growth companies, and shortages of key technical specialists. Poor communication between employers and educators, a declining enrollment in technical programs, and a lack of understanding of the industry by youth were identified as serious industry concerns. It was found that the major issues facing the industry were employers having difficulty locating competent and qualified workers and the discrepancy between the skills being taught at the post-secondary level and those required by industry. This view was confirmed in another 1992 industry-sponsored survey that indicated that the most important factor limiting industry growth was the shortage of trained workers in key technical, scientific, professional, and managerial occupations. As a result of the information obtained from these surveys, the industry announced a commitment to sustained industrial growth and implemented immediate action to address its key human resource concerns. This commitment paved the way for a business approach to EE.

A Business Approach to Environmental Education

Business is now held accountable for environmental conditions. J. E. Newall, chairman and chief executive officer of Du Pont Canada Inc., noted three areas in corporate operations that are profoundly affected by environmental issues: First, how government regulations affect business; second, how rapidly changing technology is leading both regulations and corporate strategy; and third, how industries are responding to corporate environmentalism with new management structures (Newall 1990, 58–62). A business approach to EE employs an integrated management framework that incorporates government policy and technological advancements. It is an approach that may be successfully applied to any jurisdiction capable of establishing cooperative business relationships.

A number of variables affect the management of an effective educational program. It is important, therefore, to identify and carefully analyze the program's key elements. For the environment industry, these elements are (1) supply, or human resource issues; (2) the training and education system; and (3) continuing industry demand.

The supply element is concerned with those people entering, and presently in, the work force. This involves such human resource issues as the declining number of people entering the industry, declining enrollment in science and engineering programs, a lack of awareness of environmental career opportunities, and the lack of appropriate entry-level skills and the knowledge needed for more advanced education. The challenges for the industry are to attract high-quality people as they enter the work force, encourage current practitioners to upgrade their skills and knowledge, and to attract people from "sunset" jobs.

Training and education for the industry involves everything from K through 12, college, technical institute, university, commercial, and in-house programs. Training and education issues involve the lack of recognized common standards, inconsistent approaches to the environmental sciences, and ineffective linkages with industry. Industry concerns also include funding constraints, the lack of appropriate staff, institutional inflexibility, training that may not meet industry needs, and barriers to inter-regional transfer.

The challenges are to develop and implement a consistent set of high standards and to develop programs that meet industry's needs by defining pathways for career development, retraining, and upgrading.

Finally, there is the demand—from the industry itself, other industries, government agencies, and the educational system. The key issues here include the industry's economic growth and the lack of information concerning the industry's human resources. The lack of organization and fragmentation within the industry, the limited number of established career paths, and the high rate of technological and regulatory change also have a profound effect on the ability to understand the critical skills and knowledge required for world-class environmental capability.

Putting the Business Approach to Work

In order to apply a business approach to EE successfully, critical issues must be addressed. A systematic needs analysis is required. Solid data are also required to provide a rationale and framework to gain cooperation. An understanding of the relationships among industry, government, and the public makes it easier to substantiate programming proposals. It is essential not only to provide information but also to participate in the exchange of ideas. For cooperation to result, stakeholders need opportunities for input. Agencies with vested interests must also see the financial importance of participating. In business, this amounts to a clear sense of the potential for profit. Individuals have to be told exactly how they will gain from their investment in education.

The best way to elaborate on the ESAA's business approach to EE is to provide examples of the programming initiatives currently taking place. Through the use of appropriate diagnostic measures and needs assessments, the ESAA has established an educational agenda that focuses on issues in six categories: professional development, marketing, standards, curriculum, delivery, and services.

Professional Development

Professional development is vital to the ESAA. Its annual Environment Business (EB) professional development program consists of a week-long series of courses held at the University of Alberta, ranging from one to three days in length. The ESAA's reputation for quality and integrity has gained institutional status recognition by the Canadian government that allows a tax credit for individuals attending EB courses. It is important to note that the EB offerings are usage courses and are not to be confused with the information pieces or seminars also provided by the ESAA. Environment Business tuition revenues generate a substantial portion of the funding base for the ESAA's educational program. It is for this reason that a great deal of effort goes into providing the highest quality learning experience possible, with instructional staff composed of government, academic, and industry practitioners. The excellence of the program extends beyond the industry, inasmuch as over 60 percent of the participants are non-ESAA members.

Marketing

The need for additional marketing skills is a persistent human resource issue in the environment industry. The ESAA has initiated marketing courses as part of its annual Environment Business program. However, sparse attendance indicates that more needs to be done. In this regard, the ESAA has proposed a program to train marketing professionals for the environment industry. The program, consisting of evening and weekend study, will provide training that will enable marketing professionals to make lateral career moves into the environment industry. Such professionals, armed with the technical expertise provided by the training program, will have the ability and talent to make a dramatic impact in the area of corporate culture.

Standards and Credentialing

The industry recognizes the need for standards and is actively responding through major initiatives involving the identification of needed skills and the provision of definitions for environmental work that will reduce the gap between educational qualifications and industry requirements. Nevertheless, a number of problems appear to have no clear solution. There is a great deal of difficulty in determining the purpose of credentials. The excessive number of credentials and providers, including redundant credentials, as well as overly-specialized activities, can lead to fabricated sub-discipline credentials. This, in turn, prompts the development of still more credentials that dilute their recognition value. Clearly, the profusion of "environmental" credentials provides the valuable lesson that "more isn't necessarily better."

The Federation of Environmental Professionals has identified more than 90 professional credentials being issued by 60 providers. According to surveys, a substantial majority of environmental professionals favor a single credential modeled after a professional engineer-type program. It is for this reason that the ESAA supports the Qualified Environmental Professional (QEP) designation. The QEP was initiated in 1990 as a result of overwhelming support by the environment industry for a general, multi-disciplinary credential. Briefly, the QEP designation establishes a standard for experienced professionals and provides a career track for those new to the environment industry.[3] The ESAA, as a benefit to its members, covered the cost of the application and oral examination for the initial QEP applicants. A provincial QEP chapter is now established and, at the present time, there are six QEP designation holders in Alberta, more than in any other province in Canada.

Finally, on the issue of standards, a challenging project has been suggested to the ESAA by its members to judge the "value" of different certificates and diplomas. The industry, it was suggested, may find value in "ranking" the various training institutions and the associated relative value of their degrees, certificates, and diplomas. The ESAA, however, does not desire to become a certifying body or the issuer of "seals of approval." In general, reviews (whether they be of past lovers, restaurants, or universities) tend to be inherently flawed by the lack of objective rating instruments.

Curriculum

Quality curriculum provides the nexus for the ESAA's educational programming initiatives. Training and courses must provide value to the ESAA membership, and the ESAA endeavors to offer professional development courses that are timely and of sufficient rigor and quality to meet current industry needs and standards.

The ESAA strongly believes in strengthening the communication process between education and business. Current provincial government directives, as indicated by the Alberta government's policy white paper, encourages joint education and industry partnerships. The ESAA's Education Committee, made up of industry, post-secondary education, and government leaders, provides input concerning the direction of the ESAA's educational program and strengthens the linkages between the major stakeholders.

Cooperative education, in the form of practicum and employment placements, is an integral component of many of the ESAA's educational programs. The Applied Environmental Management Program, for example, offered in conjunction with the Northern and Southern Alberta Institutes of Technology, provides retraining for displaced science and technology professionals to assist them in re–entering the workforce in the environment industry. The program includes eight weeks of classroom training and an eight-week industry practicum. The training supplements the technical and science skills this target group already possesses and is provided in environmental technology, regulations, and management. Now running for the fifth time, over 70 percent of the program's graduates work full-time within the industry.

Another strong example of cooperative education is found in the ESAA's National Youth Internship Program. The goal of this project is to produce highly competent personnel able to meet the health, safety, and environmental protection needs of small and medium-sized businesses. The program is a three-year pilot project designed to put high school graduates into career paths that progressively increase their skills, training, and academic experience. The program provides students with "cutting edge" industry-related experience and recognized academic training. It consists of 16 weeks of classroom instruction and 34 weeks of relevant employment for each student. Employers act as training hosts for the practicum portion and also provide funding support over the three years of the program.

Additional ESAA programs include the following:

- course offerings that address the needs of small firms
- a calendar of environmental courses offered in western Canada
- short courses in management through seminars and EB offerings
- nationally recognized health, safety, and environmental technician certification
- a training program for international trade officers
- alternative delivery methods for training
- electronic information services

Delivery

The ESAA makes every effort to develop programs that reflect industry needs. Flexibility with respect to program delivery is limited at this time. Traditional methods, such as classroom instructed courses, seminars, and workshops, are commonly used. The costs associated with developing distance learning programs are high, but this area is being pursued. Securing partnerships with post-secondary institutions and the use of existing technology will help a great deal. Access to educational programs is also a challenge since the seasonal nature of the industry precludes offering extensive programming at certain times of the year. The winter months are the best time to provide courses.

Services

How companies are to receive recognition for investing in employee education is open to conjecture. The use of wage subsidization is already happening in some areas such as the Youth Internship Program. However, the reward for companies investing in employee education exists intrinsically. It simply makes "good sense" for companies to encourage professional development. The ESAA's main service is to ensure that its members' views and concerns are recognized.

Making It Work

The interface between institutions and industry is well established through the ESAA's assorted focus groups and committees. Educational costs are reduced by sharing curriculum with post-secondary institutions and joint funding initiatives. Programming is a cooperative effort and involves the major stakeholders. The ESAA provides an experienced structure by which government agencies can work with post-secondary education and industry to provide educational programs in a cost-effective and efficient manner. Meeting the educational needs of industry is an ongoing and dynamic process. The synergy created through the ESAA ensures that industry, both collectively and through individual companies, has a voice through which concerns and needs are acknowledged.

Conclusions

Environmental education is a collective responsibility. Educational programs must meet currently defined needs and prepare industry for the continuing assessment of additional needs as they arise. Effective EE involves cooperation and shared financial obligation. The ESAA's training programs make tax*payers* out of tax*eaters* by providing opportunities for training to individuals ranging from young adults to seasoned industry veterans. The growth and development of the environment industry depends on a commitment to environment-oriented education. The ESAA, through careful planning and organization, demonstrates that a collection of small to medium-sized businesses can address its human resource needs and implement environmental programs that can make a difference.

Notes

1. According to Industry Canada, a small business is one with less than 50 employees and less than $5 million in annual sales. A medium-sized business has 50 to 500 employees and between $5 million and $50 million in annual sales.
2. The ESAA was originally established in 1987 as the Alberta Special Waste Services Association (ASWSA).
3. The Institute of Professional Environmental Practices (IPEP), established in 1993, is the certifying body for the QEP.

References

The Canadian Council for Human Resources in the Environment Industry. 1994. *Report 1: Definition of Environmental Employment.* Calgary: CCHREI.

The Canadian Council for Human Resources in the Environment Industry. 1994. *Report 2. Classification of Environmental Occupations.* Calgary: CCHREI.

Ernst & Young. 1992. Steering Committee of the Environment Industry. *Human Resources in the Environment Industry Detailed Report.* Ottawa: Employment and Immigration Canada, Labour Market Outlook and Sectoral Analysis Liaison Group.

Industry Canada. 1994. *Industry Association Research project; Vol. 3: Benchmarks in Innovation: Association Best Practices in Organizational Design and Service Delivery.* Ottawa: (March).

Newall, J.E. 1990. "Faulty Regulations Thwart Environmental Protection," *Issues 4,* 5 (August/September), 58-62.

Royal Bank of Canada. 1995. " Staying in School," *Royal Bank Letter 75,* 7 (January/February).

Special Waste National Adjustment (IAS) Committee. 1992. *Together into the Future* Edmonton: The Environmental Services Association of Alberta.

Toronto Globe and Mail. 1995. (February 11).

Chapter 13

Reporting Risks and Setting the Environmental Agenda: The Journalist's Responsibility

David B. Sachsman
University of Tennessee at Chattanooga

W e are three years away from the 21st century, four years away from 2001, when science fiction dreams were supposed to become realities. The 21st century was to be a time of spaceships run by thinking computers, where everything was either new, newer, or newest. But the world outside this writer's Chattanooga window is a 20th-century parking lot, and in New York City, the view from one's window and the sound from the streets (see Bronzaft, this volume) may be even worse. We have spoiled our air and our water, and the only way we have found to solve these problems is to move away from them. While we watch, the industrial age is becoming the information age—but we have not cleaned up the mess left over from the 20th century.

Evolution of Environmental Journalism

Forty years ago, we Americans were so in love with the products of our postwar economy (the automobiles, TV sets, and tract homes full of furniture) and so afraid of the Communists (jet planes, atomic bombs) that we did not think about the environment. Rachel Carson tried to wake us up with *Silent Spring* (see Corcoran, this volume) in 1962, and she did manage to open the eyes of those who would become the environmental activists of that generation. But most people remained focused on other things (jobs, weapons) until the dramatic Santa Barbara Channel-Union Oil leak of 1969. In that year, Americans stared at their TV sets in disbelief, watching sobbing college coeds wiping the muck from oil-soaked birds trapped in the spill.

Until 1969, the scientists had been the heroes, responsible for transistors, antibiotics, fast cars, color TV sets, Princess phones, stereos, and all the rest—the great inventions and technological gadgets of the modern era. If roads were clogged, so what? We could build more roads. Americans could build more roads and fight the war in Vietnam at the same time. Americans would not realize that they had run out of money until the early 1970s.

The American public was shocked in 1969, when the high-tech oil companies and their superscientists failed to prevent the oil leak that was killing all those birds right there on color TV night after night after night. The American people and the American mass media had awakened to environmental issues. In 1969, *Time* and *Saturday Review* began regular sections on the environment, *Life* and *Look* increased their coverage, and *National Geographic* offered 9,000 words of environmental coverage in a single issue. In 1970, Walter Cronkite seemed almost as interested in the environment as he was in space travel, and Paul Ehrlich's *Population Bomb* made the best-seller list.

The environment was clearly on the public agenda and the media agenda in 1969, but which came first? Did the media set the environmental agenda by carrying all those pictures of oil-soaked birds? Did the environmental activists set the agenda? Did the federal government set the agenda? Did the public itself first set the environmental agenda, and did the media just follow public interest? The one thing we know for sure is that industry and industrial scientists, who had set the environmental agenda prior to the 1960s, were now the people being blamed for oil spills and other environmental problems.

Environmental Courage: A Challenge for Journalists

Most reporters covering the environment in 1969 and 1970 were new to the beat. Some were science and health reporters. But many did not have a science background. The science reporters tended to conceptualize the environment as a scientific topic. The others often viewed it from a political, social, economic, or even esthetic perspective.

Like science writers, the science-writing teachers in journalism schools would tend to think of environmental reporting as an important new science beat, missing the fact that many environmental reporters thought of the two as separate and different. These two approaches toward environmental reporting continue to the present day. The journalists who cover the yearly convention of the American Association for the Advancement of

Science (AAAS) are not necessarily the same people who attend the annual meeting of the Society of Environmental Journalists (SEJ). The difference between science writers and everyone else who covers the environment is understood within the newsroom, where local-beat reporters and general-assignment reporters are often sent to cover such environmental stories as chemical-truck spills, zoning disputes, and sewerage board meetings (which a science writer might call "accident," "government," or "meeting" stories).

There is one group that is absolutely certain that the environment is a scientific issue—that is, the scientists themselves, whether they work for government, industry, universities, or activist organizations. While scientists acknowledge that the environment can be viewed from a political, social, economic, or perhaps even an esthetic perspective, it remains a scientific issue. And scientists wish to be judged by the standards of science rather than esthetic standards or the values of journalists. Scientists measure their work by its importance to the world, while most journalists have adopted the news values of their profession: timeliness, proximity, prominence, consequence (importance), and human interest (MacDougall 1977, 56). To journalists, it made sense to increase coverage of the environment because of the Santa Barbara Channel-Union Oil spill (which had timeliness, proximity, and human interest, as well as importance). To scientists, timeliness, proximity, prominence, and human interest are just not important (Sachsman 1994, 945).

Whose responsibility is it to inform the public about environmental issues? Is it the duty of scientists, journalists, government officials, educators, corporate leaders, environmental activists, or all of the above? Since the late 1960s, government officials have considered the environment a government story and have regularly supplied the media with more press releases and other forms of environmental information than has any other kind of source. Institutions, such as universities, also issue press releases and hold press conferences to try to reach the general public and special publics, such as government and foundation officials, through the press.

In the postwar industrial boom of the 1940s and 1950s, the environment was a business issue, thanks to the efforts of corporate public relations. Press releases from environmental activists countered corporate arguments in the 1960s, and from 1969 on, many have looked to citizen-action groups for information about the environment. Meanwhile, corporate officials will tell you that they have been in a public relations war since at least 1969 and that many journalists and government officials have joined the activists against them. Some now see the environment not just as a business story (see Ruhl, this volume) but also as an economic survival story, a matter of jobs and profits versus trees.

The environment has a different meaning for every group. Each has a different message for the public. Most would like to influence the public agenda, the list of concerns that people are talking about. While some groups are able to reach the public through advertising and other direct means, all are at least partly dependent on the mass media to carry their messages to the people. That is where they come into conflict with journalists, who think it is their responsibility to decide which stories to carry and what to say about them. And these reporters and editors make their decisions on the basis of news standards that (with the exception of importance) few people understand.

Government tries to set the public agenda through legislation and regulation. It uses the press to try to garner public support. Industry wants to influence government officials and the public. It speaks directly to government and uses advertising and direct public relations to reach the public. Companies still try to get their messages through the media, but many corporate officials no longer trust working journalists and turn instead to their business counterparts, the publishers and managers. Activist groups count on environmental reporters to get their stories across.

Other institutions, such as scientists at universities, are interested in informing the public through the press but seem to be speaking a different, more complicated language. Finally, there are those among the news media themselves who see their role as picking and choosing the list of issues they think people should discuss. Thus, they acknowledge that they play a role in setting the public agenda, though most argue that they perform their responsibility objectively, using their profession's news standards to make their decisions.

Given the disparity of interests and values among the scientists, government officials, corporate leaders, and activists who have taken part in the environmental information explosion that some call a public information war—let alone the differences between these news sources and journalists—how can scientists and others teach journalists what they need to know about such scientific and technological concerns as environmental risk and risk assessment?

Environmental Journalism Today

In 1985, The Hazardous Substance Management Research Center in New Jersey began funding a joint project of Rutgers University and the University of Medicine and Dentistry of New Jersey—Robert Wood Johnson Medical School for the design and implementation of a program of continuing education for print and broadcast journalists on risk assessment. Soon afterward, a second project was added to teach scientists, government officials, business leaders, and activists serving as news sources how to provide environmental-risk information to the media.

The five-year efforts of the "Rutgers group," so called by Victor Cohn (*Reporting on Risk* 1990), have provided much information useful to those who wish to use the mass media to communicate with the public about environmental and health risk issues.[1]

Here are some broad generalizations about environmental communication today, based on the conclusions of the Rutgers group:

• The media do not supply the public with enough information about environmental risk and risk assessment. When journalists cover an environmental emergency, they concentrate on the breaking story itself (the fire story, the leak story, etc.) and often do not think in terms of the need for background risk information (Sandman, Sachsman, et al. 1987, 100, 104).

• When environmental risk is reported by the media, it is more often alarming than reassuring. When journalists include risk information in breaking stories, it is the most basic information about the threat to human health (Sandman, Sachsman, et al. 1987, 100, 104).

• Risk information in the mass media comes largely from government, industry, and unattributed sources. Journalists usually do not seek out uninvolved experts (such as university scientists) for information about risk. They want their risk information to come from official sources, such as the government (Sandman, Sachsman, et al. 1987, 100, 104).

• News coverage of environmental risk "tends toward the extremes (a situation is risky or not risky) rather than quoting sources who take an intermediate or tentative position" (Sandman, Sachsman, et al. 1987, 100).

• Environmental articles are generally accurate. Inaccuracies tend to involve "missing information or context rather than errors of fact." There is little intentional bias. "The flagrant distortions or 'lies' about which sources sometimes complain are simply not a characteristic" (Sandman et al. 1987, 100).

• The environment has become such a journalistic staple that it is now a part of the daily coverage of local-beat reporters (who cover the sewerage and zoning board meetings) and general-assignment reporters (who are assigned to everything from speeches to traffic accidents). While many large newspapers employ some kind of science or environmental expert, such environmental specialists are not normally assigned to breaking stories, such as fires and accidents. Thus, most environmental stories are covered first by young reporters with little overall experience or knowledge about the scientific concept of risk. Fortunately, these beginning journalists are eager to learn about such complex and specialized areas as environmental risk (Sachsman, Sandman, et al. 1988a, 287).

• The three major TV networks appear to be paying more attention now to environmental issues than they did during the Rutgers group's study period nearly a decade ago. During those years (which included the devastating tragedy caused by the Union Carbide chemical release in Bhopal), the time devoted to environmental news was only 1.7 percent of the newscasts, an average of five stories a week across all networks (Greenberg, Sachsman, et al. 1989a, 120-21).

• Scientific risk generally is not the criterion used by the three major TV networks to determine what appears on the nightly news. Instead, the broadcasters use the traditional journalistic standards of timeliness, proximity, prominence, consequence, and human interest and the added television criterion of visual impact. Furthermore, the networks also consider such geographical factors as time, cost, and convenience when deciding exactly which environmental risks to cover (Greenberg, et al. 1989b, 269-75).

• The traditional journalistic values teach reporters to cover dramatic events instead of chronic issues. Editors often will not accept a story written about a chronic risk issue unless the reporter has found new and timely information (called a "news peg" or "hook") on which to hang the story. As the Rutgers group said: "Network evening news coverage surely tends to reinforce the public's overestimation of the health impact of acute risk events and underestimation of most chronic risk issues. The public's conception of risk is almost certainly distorted by television's focus on catastrophes and its dependence on films" (Greenberg, Sachsman, et al. 1989a, 123, 125).

• The journalists best able to cover complicated environmental risk issues are specialized environmental reporters, whether they be science writers or strictly environmental reporters. Environmental risk and risk assessment are concepts that are understood by these specialized reporters, who nevertheless tend to stick with their journalistic values (to the surprise and con-sternation of those scientists who thought that journalists would change their ways as soon as they learned about risk).

• Many smaller media do not employ such an expert environmental re-porter, and most large media have only one, not nearly enough for an issue that cuts across many beats, from the school board (asbestos) and zoning board to accidents (leaks) and fires (fumes). First-day or breaking environ-mental stories, whether they involve land-use disputes or leaking under-ground storage tanks (LUST), generally begin with local-beat and general-assignment reporters, who are sometimes young and inexperienced. (These stories are then screened by editors who, as journalism's gatekeepers, define the news by using journalism's traditional standards.) Fortunately, begin-ning reporters are ready and willing to learn about such complicated scien-tific issues as environmental risk and risk assessment.

Improving Environmental Reporting

The goal of the Rutgers group's Environmental Risk Reporting Project was to design and implement a program of continuing education for journal-ists about risk. The researchers studied the feasibility of using various kinds of educational techniques and devices for providing reporters and editors with information about risk and developed three initial positive recom-mendations:

1. Funding should be sought for a program to provide environmental-risk libraries for the offices of newspapers and broadcast stations.
2. An environmental-risk press kit should be designed, tested, and distrib-uted nationally.
3. Continuing education programs on environmental risk for reporters and editors should be offered throughout the country (Sandman, Sachsman, et al. 1987, 106).

The group spent several years designing model seminars and workshops and producing the prototypes for related teaching materials, including a videotape *Covering an Environmental Accident* (1985) and an *Environmen-tal Reporter's Handbook* (Sachsman et al. 1988b), which developed out of the recommendation for an environmental press kit. At the same time, the researchers were engaged in their second project on risk communication for environmental news sources, working with scientists, technicians, corporate officers, regulatory officials, and activists whose responsibility it is to inform the public through the mass media about environmental risk. In this regard, the researchers concluded: "The more productive approach in the long run may be to teach news sources to better understand and deal more effectively with the media....Environmental news sources who empathize with journalists and are willing to teach reporters about their specific fields can help make mass media coverage of environmental risk as accurate and professional as the American public deserves" (Sachsman, Sandman, et al. 1988a, 295-96).

Given the goal of communicating environmental issues in the 21st century, it is the responsibility of the scientific and educational community, government, the business and corporate world, and the activist groups concerned with public policy about the environment to teach journalists and the public about environmental risk. This is also the responsibility of those environmental reporters who are already experts (such as those who are members of the Society of Environmental Journalists). Effective educational models and prototypes are available for the development of continuing education programs for—first and foremost—the beginning reporters who are on the front line of environmental reporting, the journalism students who soon will be, and their editors, who are journalism's environmental gatekeepers.

Beginning reporters are so aware of their deficiencies in this specialized area that they will come out for an educational program on environmental and health risks. Students, on the other hand, may be dreaming of the White House beat rather than the local sewerage authority, and so environmental-risk information needs to be built into their current courses or into a lecture series where attendance is required. Editors will listen if educational programs are brought to the state, regional, and national conferences that they regularly attend. Editors, and everyone else for that matter, will learn more and stay longer if the program is entertaining and involving—a hypothetical exercise, for example, rather than a talking head.

The Rutgers group ran hypotheticals (loosely based on Fred Friendly's Columbia University programs for public television) for all kinds of journalism groups. The researchers brought along their own panel of scientific experts who had been rehearsed in how to act out each stage of the preplanned hypothetical incident. For example, the hypothetical that was enacted for the videotape *Covering an Environmental Accident* involved a chemical truck spill. It was the job of a second panel of journalists to question the scientists playing the various roles (from police officer to company spokesperson) so that the journalists could figure out what should be put in a news story. For smaller groups, a press conference format was sometimes used, with all the reporters in the audience questioning the panel of role-playing experts.

Continuing education programs, such as those developed by the Environmental Risk Reporting Project, are capable of playing an important role in helping reporters and editors get a handle on such complicated scientific topics as risk and risk assessment. They should be offered as part of the journalists' own state, regional, and national conferences. They should be conducted by journalism schools across the nation. They should be run by scientific corporations and by their industry groups. They should be produced by the various environmental-activist organizations. They should also be conducted by university-based scientists under the auspices of their institutions' continuing education programs. All of these people and organizations should take responsibility for teaching journalists about risk, and they all are able to support continuing education programs on the environment.

The members of the Rutgers group felt that every newsroom should have a shelf of basic scientific and environmental reference books (right next to the dictionary) so that reporters covering a breaking story, such as a chemical spill, could look up the chemical. As part of their continuing education programs, they sometimes handed out such books, but their attention soon turned to the creation of an environmental risk press kit, a reference work on environmental risk specifically designed for reporters, mentioned briefly above.

This environmental risk press kit grew into the *Environmental Reporter's Handbook*, a 175-page book containing 27 environmental risk briefings (from "Acid Rain" to "Wetlands"), a long list of environmental news sources (complete with addresses and telephone numbers), a section of suggested references and source books, a glossary of technical and scientific terms, and the alphabet soup of acronyms (from AEA to WHO) facing environmental journalists. It was a prototype of the kind of handbook the group felt should be produced and distributed to newsrooms every two or three years so that both the scientific information and the phone numbers would be up-to-date.

On the same track are the handbooks produced by the National Safety Council's Environmental Health Center, such as *Covering the Coasts: A*

Reporter's Guide to Coastal and Marine Resources and *Reporting on Municipal Solid Waste: A Local Issue* (1993), and Robert Logan's *Environmental Issues for the '90s: A Handbook for Journalists,* which was published by the Media Institute and the Radio and Television News Directors Foundation in 1992. The concept of a reporter's handbook has been tried and tested and found to be effective.

The Rutgers videotape *Covering an Environmental Accident* was distributed to journalism schools that promised to keep a short-term record of when and how it was used. Such educational videotapes are produced on a variety of topics and can be quite useful, but they cannot stand alone. They need to be part of a classroom setting, with a teacher who can explain things, or part of a continuing education program. And they are never as good as a live panel discussion or a live enactment of a hypothetical incident. It is very difficult to interact with a videotape. In the long run, tapes (such as *Covering an Environmental Accident*) should be replaced by interactive computer games, with the real-life reporters asking questions of the computer's scientific news sources in order to assess whether the chemical spill in the computer's hypothetical incident represents a serious risk to the public's health or wellbeing.

Setting the Environmental Agenda

The Environmental Risk Reporting Project found that it was possible for scientists and others to teach journalists what they need to know so that they can inform the public about environmental risk. The key to communication, the Rutgers group learned, was explaining each story to journalists by using commonly understood words (instead of scientific jargon) and presenting the story in terms of the journalists' own news standards of timeliness, proximity, prominence, consequence, and human interest. It has taken a long time for many news sources to realize that they cannot change the values of journalism to fit the beliefs of other fields, such as government, business, the law, or science. This is especially true about science, since the only thing scientists have in common with reporters and editors is the standard of importance.

Journalists have been able to keep control of the news—and to influence the public agenda—because they have controlled the definition of news. They have adjusted their standards only in terms of the TV criterion of visual impact, although they have also adjusted their behavior in response to the geographical and financial factors of time, cost, and convenience.

The so-called "power of the press" may very well be the power to influence the public agenda, the power to rank-order the list of topics of interest to the people. Journalists never speak of themselves in terms of power but rather in terms of their responsibility to the public. They define their responsibility in terms of their own professional value system, much as lawyers and government officials do. Other institutions and groups have the power, independent of the editorial control of the press, to influence the public agenda. Government, industry, activist groups, and educational and research institutions can send their messages directly to the people through the mail, advertisements, or personal appearances. These news sources also have the power to use the media to get their messages across. When they write their press releases to fit the rules of the game—complete with timely, local news pegs of interest to general readers—and when they stage their news events in accordance with the same criteria, their stories usually get through the gatekeepers of the press, and they can influence the public agenda just as easily by using the mail.

Given the intense interest of government officials, business executives, activists, and scientists in influencing people, who really sets the public agenda? Do the interest-group news sources or the media themselves tell people what they should be thinking about, if not what they should be thinking? The answer lies in the combination of influences that affect every news story, a combination that includes what reporters and editors bring to the story, press releases and phone calls received from a variety of news sources, and, finally, the nature of the story itself.

The important and dramatic occurrences at Three Mile Island, Bhopal, and Chernobyl set the public's environmental agenda from the late 1970s through the 1980s, just as the Santa Barbara oil spill first triggered that awareness in 1969. The very nature of the news itself has had and can have a profound effect on the public agenda. But people do not react directly to an event. They react to the images and messages they receive, images and messages that come from the combined efforts of news sources and journalists. They react not just to the Santa Barbara oil spill but also to the televised images of tear-filled college students with dying birds in their arms. They react not just to the problems at Three Mile Island but also to the warnings of possible danger. And where the facts were truly horrifying—at Bhopal and Chernobyl—the world reacted to the media images and still probably does not have a true sense of the reality of those tragedies.

The combination of sources, gatekeepers, and events creates the stories and images that influence public thinking. The mix of forces changes every day with every story. Sometimes business controls the message; sometimes activists or the nature of the story wins out. Often the media themselves determine which pictures and images are seen by the public. The journalists clearly have the advantage. They not only write the story; they are the gatekeepers who decide whether the story is sent on to the public, and they make their decisions on the basis of rules of their own creation, the standards of their profession. Anyone wishing to influence the press must learn to play by those rules.

Note

1. The group consists of Dr. Michael R. Greenberg, professor of urban planning and public health; Dr. Peter M. Sandman, Professor of Environmental Journalism and Director, Environmental Communication Research Program, Rutgers University; and David B. Sachsman, Dean, School of Communications, California State University at Fullerton. For convenience, I have referred to them in several instances as "the Rutgers group" or "the Rutgers professors," since Sachsman is a former New Jerseyite (Cohn 1990, 61).

References

Andrews, John S., et al., eds. 1994. *Hazardous Waste and Public Health: International Congress on the Health Effects of Hazardous Waste.* Princeton: Princeton Scientific Publishing Co.

Cohn, V. 1990. *Reporting on Risk.* Washington, D.C.: The Media Institute.

Greenberg, M.R. & D.B. Sachsman, et al. 1989a. Network evening news coverage of environmental risk. *Risk Analysis 9,* 120-21, 123, 125.

Greenberg, M.R., & D.B. Sachsman, et al. 1989b. "Risk Drama and Geography in Coverage of Environmental Risk by Network TV." *Journalism Quarterly 66,* 269-75.

Logan, R.A. with W. Gibbons & S. Kingsbury. 1992. *Environmental Issues for the '90s: A Handbook for Journalists.* Washington, D.C.: The Media Institute and the Radio and Television News Directors Foundation.

MacDougall, C. 1977. *Interpretative Reporting, 7th ed.* New York: Macmillan.

National Safety Council Environmental Health Center. *Covering the Coasts: A Reporter's Guide to Coastal and Marine Resources.* Washington, D.C.

National Safety Council Environmental Health Center. 1993. *Reporting on Municipal Solid Waste: A Local Issue.* Washington, D.C.

Sachsman, D.B., producer. 1985. "Covering an Environmental Accident." Distributed by the Environmental Communication Research Program, Cook College, Rutgers University, New Brunswick, NJ.

Sachsman, D.B. 1994. "Communication between scientists and the media: Introducing the concepts of risk, risk analysis, and risk communication to journalists." In Andrews et al. (above).

Sachsman, D.B., et al. 1988b. *Environmental Reporter's Handbook.* Newark, NJ: Environmental Risk Reporting Project, Hazardous Substance Management Research Center, New Jersey Institute of Technology.

Sachsman, D.B., et al. 1988a. "Improving Press Coverage of Environmental Risk." *Industrial Crisis Quarterly 2,* 287, 295-296.

Sandman P.M., D.B. Sachsman, et al. 1987. *Environmental Risk and the Press: An Exploratory Assessment.* New Brunswick, NJ.: Transaction Inc., 100, 104, 106.

Chapter 14

Writing the Wilderness: The Challenge of "Complex Pastoralism"

Gene McQuillan
Kingsborough Community College
The City University of New York

In 1989, I went with a group of experienced cavers on an excursion to West Virginia, which features spectacular limestone caves, several of which have not been fully explored. Many of the entrances to these caves are on private property, and because of concerns about overuse, vandalism, wildlife preservation, and potential lawsuits, many landowners have restricted or completely denied access to them. One owner made his point with a large sign:

IS THERE LIFE AFTER DEATH?
STEP ON MY PROPERTY
AND YOU'LL FIND OUT!

One might read this sort of sign as an obituary for nature writing in modern America. This chapter, however, does not focus on bitter responses to land that has been grabbed and locked away by land speculators, logging companies, or other villains of the conservation movement; it focuses instead on the intense ambivalence that writers have displayed toward lands which have been "saved." In simple terms, American nature writers have yet to find a way to respond to a second type of sign:

YOU ARE NOW ENTERING
GLACIER PEAK WILDERNESS AREA

It is hardly surprising that a writer would react with anger or disillusion to the sight of an owner's warning. One's reactions might be complicated if the "fence"—be it barbed wire, a sign, a treaty line, or a national park boundary—existed because of the direct or indirect actions of the author. Numerous texts about American wilderness—most notably Thoreau's "Walking," Edward Abbey's *Desert Solitaire*, Paul Schullery's *Mountain Time*, Aldo Leopold's "Land Pathology," and John McPhee's *Encounters with the Archdruid*—confront a bitter necessity, namely that the continued preservation of American wilderness has depended on the public ownership of land. Whether it is a town's land grants, an area purchased by the Nature Conservancy, or a national park, American writers have often participated in efforts toward the ownership of wilderness, including Thoreau's calls for town-owned land preserves, John Muir's fight for Hetch Hetchy Valley, Edward Abbey's experience as a park ranger at Arches National Monument, and Anne LaBastille's support for Adirondack State Park. While writers have often fought for such methods of wilderness preservation, they have at times found that their own will to write about wilderness was seriously undermined by the very actions they had supported.

Our Vanishing Wilderness

As the wilderness areas of America continued to disappear throughout the 19th century, American authors began to intensify their questions about the relation of wilderness to ownership. In his essay "Walking," Henry David Thoreau asks a question that would become a crucial theme in American nature writing: "What business have I in the woods, if I am thinking of something out of the woods? I suspect myself, and cannot help a shudder, when I find myself so implicated" (1981, 377). Yet it is also important to note that Thoreau's attempts to release himself from the confines of fences and lots are not often figured as trips to uncharted regions. Even in a passage from his *Journals* (21 July 1851) where Thoreau states "my spirit is free," the image of the road still signifies Thoreau's attachments:

> Now I yearn for one of those old meandering dry uninhabited roads which lead away from towns—which lead us away from temptation, which conduct to the outside of the earth…where you can forget what country you are traveling…where travelers are not too often to be met— where my spirit is free—where your head is more in heaven than your feet are on earth—which have long reaches—where earth is cheap enough by being public. Where you can walk and think with least obstruction. (1949, Vol. II, 322-23)

This description of a road with the "least obstruction" actually becomes a catalog of such intrusions: temptation, geography, economics, leisure, other travelers, and, of course, towns. Thoreau states that such roads serve as a temporary refuge "where you can forget what country you are traveling," but the "pilgrim" who wrote this passage repeatedly reminds us of his allegiances. He writes in the same entry, "In roads the obstructions are not under my feet." Such claims strike one as an evasion of the actual question of land ownership, of how the diminishing possibilities of writing about wilderness and landscape are indeed substantial and urgent claims upon his attention and awareness. In this passage, one might assume that Thoreau associates himself with the "uninhabited roads," not with "the town," "the farmer," or "the gentleman." One who pictures Thoreau striding along "where walls and fences are not cared for....where earth is cheap enough by being public" should also remember that Thoreau earned most of his money during this time by surveying the lands of the town, the farmer, and the gentleman.

It is tempting to see Thoreau's surveying not only as compromise but also as failure: "I suspect myself and cannot help a shudder when I find myself so implicated." Yet it is this problem of "implication" that grounds Thoreau in a series of considerations which become much more convincing and relevant than the naive and almost random suggestions of "Walking." While Thoreau continued to sense a strong and basic opposition between the town and the woods, he occasionally suggested that the woods be saved or managed by the very township which threatened their destruction: "In the mid-1850's he made a proposal that each town in Massachusetts save a 500-acre tract of woods which would be forever wild: no lumbering or changes at all" (Perrin 1987, 20). Instead of continued pleas for "absolute freedom and wildness," Thoreau adopted a position that would become more prevalent in the latter years of the 19th century: in his latest and soberest years, he became, to some extent, that very tame and civilized thing, "a conservationist," and he even made a plea for the establishment of national parks (Krutch 1965, 55).

For recent nature writers accustomed, devoted—or perhaps even resigned to—the idea of national parks, Thoreau's surveying provides a striking image of ambivalence and frustration:

> The history of woodlots is often...a history of cross purposes—of steady
> and consistent endeavor on the part of Nature—of interference and

blundering with a glimmer of intelligence at the 11th hour on the part of the proprietor…What shall we say to that management that halts between two courses? Does neither this nor that but botches both? (1983, 396)

Thoreau may not have provided the solutions to the problem of ownership, but he has provided some key terms for recognizing its effects. These "cross purposes" of "interference" and "management" are the legacy of a 20th-century wilderness area, whose proprietor's general plan for easy tourist access in pristine wilderness areas certainly "does neither this nor that but botches both." The "eleventh-hour" urgency with which people preserve wild lands or encourage environmental policies leaves many "woodlots" in a precarious state. Thoreau had warned of "the evil days," and by the end of the 19th century, there were growing numbers of people who were willing to accept this prophecy. Some saw an alternative, a way to forestall the abuse of the land. America had lost Eden but still possessed a Yellowstone, a Yosemite, a Grand Canyon. The questions were, who would own them, and would this ownership be yet another curse upon the land?

"Interference and Blundering with a Glimmer of Intelligence"

The history of the national parks is long and varied and has already received the attention of numerous scholars and historians. There has been a minor industry of books, ranging from coffee-table photo collections to more serious meditations on the status of wilderness as it exists in the national parks. Yet there has been a gap, marked by silence, apathy, and even open disdain, between literary communities and the national parks. In simple terms, the establishment of these parks, which seemed to be the most likely means of preserving American lands in a time of economic pressures, apparently offered few possibilities for the modern writer. A harsh, accusatory tone often marks the commentary about the parks, which are often associated with various types of over-crowding and tourism, with "the well-equipped Republican clones you see marching like Hitler youth up and down the spine of the Rockies, or in any of the national parks, national forests, wilderness areas in America" (Harrison 1987, 230). Those authors who chose to write about and within the park system soon came to share Thoreau's fears of management based on "interference and blundering with a glimmer of intelligence."

In 1916, Congress instructed the Park Service to manage the parks and "to provide for the enjoyment of same in such manner as will leave them unimpaired for the enjoyment of future generations." As more and more people began to visit the parks, the contradiction of providing for the enjoyment of unimpaired lands became the source of bitter debate. The parks provided a well intended yet desperate means of ownership that would forestall the destruction of wilderness. As Roderick Nash explains in *Wilderness and the American Mind* (1990), "For over a century wilderness advocates labored for just such a goal. They reasoned that preserving wild places depended on getting Americans into them without saws or bulldozers, only to find in this success the source of their greatest challenge" (236). The "wilderness advocates" to whom Nash refers have often responded strongly to such contradictions, and some of the most troubling comments about the park system come from those who worked for it in some way. In *Desert Solitaire* (1968), Edward Abbey laments the changes that have happened to the wilderness and to the rangers who serve as its "advocates."

Even those writers who actively support the idea of national parks have serious doubts about their benefits. After Paul Schullery worked as a ranger/naturalist in Yellowstone, he offered this warning in *Mountain Time* (1984):

> I warn you also that the national park idea is a philosophical rat's nest, a self contradictory, ironic, incurably anomalous, and socially anachronistic bundle of challenges and frustrations. . . The American public has never received an adequate introduction to the National Park idea. To them, or most of them, the parks are little more than grassy Disneylands… National parks, forests, monuments, and other federal reservations, as well as similar state areas, are all viewed, simply, as places of recreation. (72-73)

Of course, these accusations feature more than their share of exaggeration, and readers of *Desert Solitaire* and *Mountain Time* will find much more complex and substantial assessments of the park system. Yet at the heart of this debate is the sense that the parks represent only a small step toward an ecological consciousness or a "land ethic." Aside from cheap shots at tourists and stories of gridlock in Yosemite, there have been more insightful criticisms of the contradictions that are bound to arise in a country in which a President can say that "he wanted to be remembered as the greatest conservationist and the greatest developer of all time" (Reisner 1986, 151). As Aldo Leopold stated in a speech entitled "Land Pathology" (1991):

> A few parcels of outstanding scenery are immured as parks, but under the onslaughts of mass transportation their possible function as "outdoor universities" is being impaired by the very human need which impelled their creation. Parks are overcrowded hospitals trying to cope with an epidemic of esthetic rickets…. (216)

While many nature writers would support Leopold's pleas, they still sense that national parks and similar areas will have to suffice until the "epidemic of esthetic rickets" has subsided.

Wilderness Trails (and Other Contradictions)

The attempts to preserve a reasonable balance of economic and esthetic needs are dominant themes in the writings of John McPhee. His *Encounters with the Archdruid* (1971) is a series of meetings and bitter debates between David Brower, former director of the Sierra Club, and mining engineer Charles Park, real estate developer Charles Fraser, and dam builder and Bureau of Reclamation chief Floyd Dominy. The crucial issue is whether wilderness experience and wilderness itself benefit from government regulation of lands. Part One of the book, "A Mountain," features a trip to the same Glacier Peak Wilderness Area that Snyder had previously visited. McPhee's companions for the trip are Brower and Park. The conflicts that arise in the essay are in regard to the pristine beauty of the area—and the possibility that it would soon be open to large-scale copper mining operations, despite the apparent protection given the area by law, specifically the 1964 Wilderness Act:

> In 1964, the United States Congress set aside this region and others as a permanent wilderness, not to receive even the use given a national park, not to be entered by a machine of any kind except in extreme emergency, not to be developed or altered or lumbered—forevermore. Within the structure of this so-called Wilderness Act, however, was a provision known as the "mining exception": all established claims would remain open to mining, and new claims could be made in any wilderness until 1984. (McPhee 1971, 4-5)

McPhee raises a pertinent point: To what extent does the establishment of wilderness area compromise the appreciation of wilderness by the very people who love it and defend it? The essay soon announces that this wilderness is not that of the frontier, the trapper, or the homesteader. This habitat owes at least part of its existence to the legislator and the administrator:

> Not far into the mountains. . . we had come to an oddly formal landmark. It was a sign that said, "You Are Now Entering Glacier Peak Wilderness Area." In other words, "Take one more step and, by decree, you will enter a preserved and separate world, you will pass from civilization into wilderness." Wilderness was now that demonstrable, and could be entered in the sense that one enters a room. (7)

The sense of wilderness as a construct, as an area to be "established" or "set aside" by the same Congress that had supported the logging of nearby old-growth forests, continues to gnaw at McPhee and his companions. They face two seemingly divergent sources of "interference." The first would be the potential damage left by copper mining, which would be like "hitting a pretty girl in the face with a shovel." Yet the confrontation facing McPhee, Brower, and Park is not merely with the copper company, but with the trail system of the park, and the three hikers are quite open about their confused reactions to it. At separate points, McPhee bluntly asks Brower and Park if this wilderness area is indeed wilderness:

> I asked him [Brower] if, by his own standards, he would describe the terrain we were in as wilderness. "Yes, it is wilderness," he said. "The Sierra is what I love, but these mountains are perhaps the most beautiful we have." (14)

> Having moved above the trees into a clear area, Park stopped to look back over the forest, the green lakes, the glacier, the snowfields, and the white peaks beyond. I asked him if, from his experience, he would call this wilderness. "No," he said. "Not with this trail in it." (17)

The most confusing moment for the group occurs at Image Lake.[1] McPhee notes that the remaining possibilities for "wilderness experience" are tainted by the consumer culture that has recently sprouted to support the recreational aspects of the outdoors and by the trail system which efficiently leads hordes of people to a stylized encounter with rugged peaks and alpine scenes. The existence of Glacier Peak Wilderness Area is the direct result of the ideas and efforts of writers such as Catlin, Thoreau, Muir,

Leopold, and Abbey. Yet what would they say about this description of Image Lake?

> Image Lake is very small—a stock water pond in size—and it stands in an open and almost treeless terrain. Slowly, we went around it, looking for a place to sleep. The sun was setting, and we had arrived much too late. We walked past tents along the shore-blue tents, green tents, red tents, orange tents. The evening air was so still we could hear voices all around the lake. We heard transistor radios. People greeted us as we went by…the lake that night had the ambiance of a cold and crowded oasis. Shivering, I climbed up the slope to witness in the water the fading image of the great mountain. Objectively, the reflection was all it was said to be. But a "No Vacancy" sign seemed to hang in the air over the lake. A real sign pointed the way to a privy. We collected firewood, which was very hard to find, and when we had something of a blaze going and had all drawn in close around it for warmth, I said to Park and Brower, "Do you feel that you're in a wilderness now?" (58)

When one first reads Park's response that such a large area loses its status as wilderness because of "this trail in it," it might seem that he is quibbling about an inevitable aspect of almost any wilderness area; Park certainly doesn't seem eager to strike off through untrailed forests or scree fields. Yet all three men share McPhee's foreboding that "we had arrived much too late." Of course, McPhee's major concern is not that they had to choose a less than ideal campsite—in a broader sense, they had arrived "too late" to see any wilderness in America.

One need not be a cynic to be disappointed by the limited and contrived prospects for wilderness experience in modern America, but one should also realize that an increased perception of disappointment, conflict, contradiction, or "artificiality" does not invalidate modern wilderness writing. Rather, the inclusion and open admittance of such tensions is a key element of "complex pastoralism," which continues to define the shifting relations of wilderness to ownership as an urgent concern of nature writers. Understanding the texts of modern nature writers requires that readers revise or set aside some of the expectations that they might have acquired from reading 19th-century texts. Writing about the wilderness may initially seem to offer the prospect of "absolute freedom and wildness," but even a casual reading of authors such as Irving, Cooper, Thoreau, or Parkman reveals the contradictory impulses that a modern reader might expect to find in texts by Peter Matthiessen, Leslie Marmon Silko, Barry Lopez, Wendell Berry, or Terry Tempest Williams. Thoreau's questions about management lead to a different construct of wilderness, one which no longer places a priority upon a strictly individual contact with the essence of nature. Instead, we are encouraged to see nature and wilderness as the product of social policies, as well as the alternative to such policies.

Note

1. Image Lake was, incidentally, chosen as one of the "Ten Best Hike-In Views" in America by *Backpacker* magazine.

References

Abbey, Edward. 1968. "Industrial Tourism." In *Desert Solitaire.* New York: Ballantine, 45-67.

Cameron, Sharon. 1985. *Writing Nature: Henry Thoreau's Journal.* New York: Oxford.

Flader, S. L. & J. B. Callicott, eds. 1991. *The River of the Mother of God and Other Essays.* Madison, WI: University of Wisconsin Press.

Halpern, Daniel, ed. 1987. *On Nature: Nature, Landscape, and Natural History.* San Francisco: North Point Press.

Harrison, Jim. 1987. "Passacaglia on Getting Lost." In Halpern (above), 230-38.

Krutch, Joseph Wood. 1965. *Henry David Thoreau.* New York: Dell Publishing.

Leopold, Aldo. 1991. "Land Pathology." In Flader & Callicott (above), 212-17.

Marx, Leo. 1964. The *Machine in the Garden: Technology and the Pastoral Ideal in America.* New York: Oxford.

_____. 1988. The *Pilot and the Passenger: Essays on Literature, Technology and Culture in the United States.* New York: Oxford.

McPhee, John. 1971. *Encounters with the Archdruid.* New York: Farrar, Straus & Giroux.

_____. 1977. *Coming into the Country.* New York: Bantam.

Nash, Roderick. 1990. *Wilderness and the American Mind.* 3rd ed. New Haven: Yale University Press.

Perrin, Noel. 1987. "Forever Virgin: The American View of Nature." In D. Halpern (above), 11-20.

Reisner, Mark. 1986. *Cadillac Desert: The American West and Its Disappearing Water.* New York: Penguin.

Schullery, Paul. 1984. *Mountain Time: Man Meets Wilderness in Yellowstone.* New York: Simon & Schuster.

"The Ten Best Trails of the World" and "The Ten Best Hike-in Views." 1988. *Backpacker: The Magazine of Wilderness Adventure 16* (January), 66-7; 52-53.

Thoreau, Henry David. 1983. "Dispersion of Forest Seeds." Unp. ms. in Berg Collection, New York Public Library. Quotations from John Hildebidle's *Thoreau: A Naturalist's Liberty.* Cambridge: Harvard University Press.

_____. 1981. "Walking." In Lily Owens, ed. *Works of Henry David Thoreau.* New York: Avenel Books.

_____. 1949. *The Journals of Henry D. Thoreau.* Vols. I-IV. Bradford Torrey & Francis J. Allern, eds. Boston: Houghton Mifflin.

Chapter 15

Environmental Images and the Luquillo Rainforest of Puerto Rico

Nancy E. Wright
Graduate School and University Center
The City University of New York

As we move toward the 21st century, it appears worthwhile to look at how different societies have managed and conserved natural resources. By understanding the different economic, political, and social interests involved in resource use over time, we may be able to enter the next century with a greater measure of wisdom regarding how best to conserve resources.

This chapter suggests that environmental concern is an expression of political culture and that political culture is a dynamic, reciprocal process that influences and is influenced by institutional and historical experience. To illustrate this approach, eight environmental images are presented to show how they reflect different aspects of use and conservation of the Luquillo Rainforest of Puerto Rico (hereafter identified as Luquillo).[1]

The Development of the Political Culture Approach

The scholars who first presented the concept of political culture based their interpretations on psychological and sociological studies of culture and individual behavior by Max Weber and Talcott Parsons. Gabriel Almond and Sidney Verba developed a comparative framework for the study of political culture, which they defined as "a particular pattern of orientations to political action" (Almond 1956). Later, Sidney Verba stated that political culture "consists of the system of empirical beliefs, expressive symbols, and values which define the situation in which political action takes place" (Pye & Verba 1965). These early political culture theorists have encountered criticism for adopting a Western-biased, one-directional perspective regarding political development (Almond & Verba 1989).

The political culture research of the 1970s and 1980s retained a certain measure of concern about transitions to democracy in developing countries; however, the research on Western industrialized countries shifted its focus from democratic stability to democratic performance and expectation. With this transition came the first references to political culture and environmental concern (Inglehart 1990; Dalton 1984). A primary strength of these political culture theorists is their identification of the significance of symbols and values in shaping political activity.

Of the vast range of environmental issues and policies that lend themselves to comparative analysis, perhaps forest use and conservation represent the greatest composite of human interests. Forests formed the very foundation of pre-agrarian, agrarian, and pre-industrial subsistence, both among the earliest tribal peoples and among peasants who depended heavily on forest products even when they did not own the land on which these products were harvested. Forests have provided the essence of human survival and have been valued for commercial purposes, including shipbuilding, fuel, and lumber. In addition, forests have held great symbolic value, serving as places for such diverse acts as refuge, meditation, and warfare.

Finally, forests have historically been places of recreation, ranging from royal hunting expeditions to family outings. Thus when we speak of forests, we speak not only of the wood or other products that forests provide but also of the land itself and a concept of place. These multiple references are important when considering politics and culture. Luquillo offers the opportunity to study change and continuity of environmental concern in a single place under three distinct types of resource management. These are pre-Columbian society, the Spanish Crown, and the U.S. Forest Service.

In the following sections, I present eight environmental images that have emerged from the management of Luquillo at various points in Puerto Rico's history. The North/South division reveals the broad differences between highly industrialized and complex societies and pre-industrial societies as well as a fundamental distinction between societies that represent former or present colonial powers and those that represent peoples who have been or are colonized.

Images of the South

Indigenous Image

The indigenous image refers to beliefs, values, and practices that comprise the relationship of indigenous peoples to their natural environments. Within the indigenous image, the natural environment directly shapes all aspects of daily life as well as cultural development. The relationship between indigenous societies and their environments is not always harmonious. However, there is often an acceptance of the nonhuman forces of nature as stronger than human capabilities, whether those forces are benevolent or malevolent. The indigenous image encompasses prehistoric, nomadic, and hunting and gathering societies as well as agrarian societies.

In Puerto Rico, the indigenous image is manifested in the Taino culture, both as it flourished in the pre-Columbian era and as it persists today. One can find examples in the symbolic value of Luquillo, which played a pivotal role in all facets of Taino life. By far the most important of these was agricultural (Pico 1986). In addition, the forest was the source of wood for canoes and later for weapons. The ceiba tree, used in making canoes, had a sacred connotation. Members of Taino society would meet in its shadow to make important decisions. Forest products held great medicinal value. Recreationally and esthetically, the forest was equally vital, supplying wood for the creation of musical instruments—the maraca, the guiro, and the drum. With the arrival of the Spaniards, the symbolic value of the forest increased; Luquillo became a refuge and a secret meeting place to plan strategies of survival and retaliation against Spanish conquest (Dominguez Cristobal 1985).

Since the 1970s, the symbolism of Luquillo has centered more on its esthetic and recreational functions. Still, elements of Taino culture persist and are reinforced by historical events. For example, some Puerto Ricans recalled the benevolent god Yuqiyu when Hurricane Hugo hit the island but did not strike its center because the high mountains created changes in wind patterns. Moreover, the rainforest is synonymous with heritage. As one former administrator with Puerto Rico's Department of Natural Resources said, "Puerto Ricans are directed to the protection of our roots, and part of our roots are in El Yunque" (Pinto 1994).

The degree to which elements of Puerto Rico's indigenous environmental image (manifested in Taino culture) remain is difficult to ascertain. In the mid-1970s anthropologist Ronald J. Duncan found that the majority of Puerto Ricans whom he interviewed did believe that Tainos exerted an influence on contemporary Puerto Rican culture (Duncan 1974). Still, it is important to note that more than 90 percent of those interviewed were students of anthropology, sociology, psychology, and/or history, and thus would be more sensitized to Taino culture. Furthermore, most who identified with Taino culture as part of their heritage could not elaborate on why they felt the way they did (ibid).

These findings suggest that some modern interpretations, rather than actual remnants, of Taino culture have found their way into the postmaterialist and biosphere images of the North discussed in the next section.

Peasant Image

The peasant environmental image is based on subsistence agriculture. Like the indigenous image, it is represented by a subsistence and largely agrarian lifestyle. Unlike the indigenous image, however, it embodies only partial self-reliance. The peasant environmental image always recognizes the possibility of displacement from natural habitats and restrictions on the use of resources such as forests. With respect to Luquillo, the peasant image is exemplified by the inhabitants of what became the Luquillo Reserve at the beginning of the 20th century, as well as by the peasant villages near the forest. For the early peasants, the forest supplied wood; for both the early peasants and today's villagers, the forest has been vital to ensuring an adequate supply of potable water.

Among the many aspects of the peasant image which warrant further research, relocation and cutting stand out. The apparent willingness to relocate distinguishes the peasant image from the indigenous image. While peasants maintain a large degree of practical attachment to the land, they may be more willing to relocate in the interest of economic improvement than adherents to the indigenous image, whose ties to land transcend purely economic considerations. Soon after conducting its survey, the U.S. Forest Service posted notices prohibiting cutting on the Luquillo reserve. The impact and extensiveness of this prohibition on the Luquillo peasant population is unclear.

Environmental Justice Image

The environmental justice and self-determination images are often closely linked, even though they differ fundamentally in their belief of who should exercise political authority over the environment. The environmental justice image is based on the premise that pollution and resource depletion are not evenly distributed and that poor and oppressed countries and sectors of society bear a disproportionate share of air and water contamination, proximity to waste disposal facilities, and an insufficient resource supply. Conversely, they also enjoy a disproportionately low share of environmental benefits, such as fertile soil, adequate supply of energy resources, and open space. At the societal level, the environmental justice image calls for political pressure to reverse the unequal distribution of environmental deterioration. At the same time, this pressure does not include a demand for political autonomy; rather, it stops with the argument that justice means equitable distribution of environmental "goods" and "bads." In this way, it differs from the self-determination image.

Self-Determination Image

The self-determination image refers to the set of beliefs that links environmental sustainablility with political and economic self-determination. According to this image, the guarantee of adequate natural resources and minimal levels of pollution in a given territory cannot be secure as long as that territory is governed by an outside authority and/or is host to outside enterprises, such as multinational corporations.

One of the strongest manifestations of both the justice and the self-determination images in Puerto Rico has been the creation by former commonwealth senatorial candidate Dr. Neftalí Garcia, founder and director of the Centro de Información, Investigación, y Educación Social (Center for Social Information, Research, and Education [CIIES]). In 1989, Dr. Garcia first proposed the formation of a broad-based Puerto Rican political movement aimed at self-determination that would include (but not be limited to) environmental protection and ecologically sustainable development. This program formed the cornerstone of his 1992 campaign for the commonwealth senate. The organization which Dr. Garcia formed following his campaign encompasses advocates of Puerto Rican independence as well as those who wish to maintain current political relations with the United States while striving for more autonomy over resource management (Garcia 1993).

Images of the North

Colonial/Mercantilist Image

The colonial/mercantilist environmental image is directed toward the insurance of an adequate natural resource supply in order to meet a state's economic and political interests at home and abroad. For this analysis, the most directly relevant example of the colonial/mercantilist image is the creation of forest codes in a number of European countries, including Spain.

In studying the effects of colonialism on developing countries, the emphasis has often been on rampant resource depletion at the hands of European powers. While this massive destruction indeed did take place, the same colonial powers also established the means for managing resources with a view to maximizing use and ensuring future supplies, especially at the first signs of depletion. The key point with respect to this type of conservation, of course, is that it was designed for the benefit of the colonizers, not the indigenous colonized.

Documentation on Spanish forest policy in Puerto Rico consists almost entirely of a record of law (Wadsworth 1949, 113ff). The earliest of the laws which I have found documented is the *Ley Primera* (First Law) of 1513, in which King Ferdinand V offered land to settlers on the condition that part of that land would be designated for tree planting. In 1536 Emperor Charles V ordered settlers to whom lands were distributed to take possession thereof within three months and plant willows and other trees along their boundaries to reinforce the timber supply (ibid, 113).

The next reference to concern about forest use appears in 1824. This document presents the complaint of Miguel de la Torre about deforestation on the island, along with the suggestion that strips of trees be left at the sources of streams. Fifteen years later the King ordered the creation of a board to protect forests, fish, and wildlife. Four years passed before the board actually approved resolutions regulating the cutting of trees. Among the key provisions was the requirement that cutting trees for sale had to be reported to the government. Cutting trees for one's own use on one's own property, however, was permitted (Wadsworth, ibid).

In 1776 the Spanish Crown ceded the lands of the Luquillo Mountains to a French duke. Spain later reclaimed them and in 1870 set them aside as one of the first legally proclaimed forest reserves of the Western hemisphere (Dominguez Cristobal 1989).

Wilderness/Wise Use Image

The wilderness/wise use image refers to the dual concerns about wilderness protection and forest management which arose in Western Europe and the United States during the mid-19th century. This image advocates what it considers a rational and balanced management of forests for economic and scientific benefit as well as for esthetic and recreational value.

In Puerto Rico, the wilderness/wise use image was introduced by the U.S. Forest Service following the transfer of the island from Spain to the United States in 1898. Despite the fact that the U.S. Forest Service declared Luquillo a forest reserve in 1903, it did not assign its first Forest Supervisor to the area until 1917. In 1922 the Forest Service established a forest nursery to produce seedlings for experimental tree plantings, and in 1939 it established its Tropical Forest Experiment Station. In 1943, the organization administering the Luquillo Reserve and the Forest Experiment Station combined to become the Institute of Tropical Forestry.

In 1916, the U.S. Forest Service, through the Commonwealth Government of Puerto Rico, completed its survey of Luquillo. Since the time of that first survey to the present, the wilderness/wise use image continues to be manifested in U.S. Forest Service administration of Luquillo, a responsibility which involves both management and scientific research.

Postmaterialist Image

This image, which stems from the traditional approaches to political culture described above, views environmental protection initiatives primarily as post-industrial responses to increased pollution and congestion and decreased natural resource supply and open space. The postmaterialist image holds that members of society who have attained a certain measure of material security are more likely to be concerned about the environment than are the poor.

The primary manifestation of the postmaterialist image in Puerto Rico is the interpretation of the indigenous image noted earlier. The attachment to Taino culture without being able to articulate the reasons for such adherence suggests a certain dissatisfaction with complex modern society, as well as with centuries of European and North American involvement in Puerto Rican life. These references to Taino culture exemplify the postmaterialist rather than the indigenous image because they are contemporary aberrations of a pre-Columbian society.

Biosphere Image

The biosphere image highlights the tension between economic development and environmental protection and seeks to resolve that tension through advocacy of sustainable development (Brundtland Commission 1987). From the 1972 United Nations Conference on the Human Environment in Stockholm to the 1992 United Nations Conference on Environment and Development in Rio de Janeiro, economists and environmentalists alike have grappled with the relationship of economic growth and environmental protection. They began this struggle by questioning whether economic growth and environmental sustainability were inimical to each other; this struggle has culminated in an effort to achieve environmentally sustainable economic development by building environmental protection measures into economic policies.

The biosphere image prescribes for the South a path of slower economic growth than that which has taken place in the North; it also prescribes deceleration in economic growth in the North. For the North, this suggests a move toward self-regulation following long periods of rapid industrialization. From much of the South's perspective, however, the advice of a slower pace of economic development also represents further attempts by the North to control and curb the South.

One important feature of the biosphere image in this regard is the identification of ecosystems, or biosphere sites, throughout the world. Specifically, the biosphere image is present in Puerto Rico's environmental protection policies through the designation in 1975 of Luquillo as a UNESCO biosphere site. This new status has resulted in an increase in the scope and diversity of environmental research and other cooperative efforts taking place in the rainforest.

Conclusions

The images presented in this analysis are merely the beginning. Refining measurement and characteristics of the images is important in answering such questions as: What is the relationship between a society's level of economic and political development and the emergence of the mercantilist image? Similarly, at what point does a society make a transition to a postmaterialist image? To what extent and in what form does the indigenous image persist in modern societies? Under what conditions are the environmental justice and self-determination images linked or separate? Do the wilderness/wise use and biosphere images encompass some of the elements of the South? What determines a transition from the indigenous to the peasant image?

Although social scientists can assess the impact of a sudden event in nature (such as a hurricane) on a given population, there is no means to measure natural changes over time. For example, what effect does ecosystem change have on a society's political choices? How can one separate ecosystem change that occurs apart from human activity from that which occurs as the result of human intervention? To what extent do members of a given society know the difference between the two, and to what extent does this knowledge affect its political culture? These are just some of the questions that call for further research on environmental concern as an expression of political culture.

The intent of this analysis is to move toward a more complex understanding of past and present environmental concern in Puerto Rico as well as among societies that have traditionally been considered ignorant, apathetic, or hostile to environmental protection. Hopefully, such an analysis can make some contribution toward the realization of environmental policies that are sensitive both to long-term resource management and to the immediate needs of all citizens, including the authority to make policy decisions about how their natural resources are managed.

Note

1. I wish to acknowledge with gratitude the support of the Intercambio Program of Hunter College, The City University of New York, which sponsored my field research. In addition, I wish to acknowledge the following individuals for their constructive comments during the research process: Prof. Christa Altenstetter, Department of Political Science, CUNY Graduate School and Queens College; Prof. Kenneth P. Erickson, Department of Political Science, CUNY Graduate School and Hunter College; Prof. Antonio Lauria, Director, Intercambio Program, Hunter College; and Ms. Arlene Dávila, Ph.D. Candidate, Department of Anthropology, Hunter College.

References

Almond, Gabriel A. 1956. "Comparative Political Systems." *Journal of Politics 13* (August), 390-409.

Almond, Gabriel & Sidney Verba, eds. 1989. *The Civic Culture Revisited.* Newbury Park, CA: Sage.

Brundtland Commission. 1987. Our *Common Future*. Cambridge: Oxford University Press.

Caribbean National Forest/Luquillo Experimental Forest, *Draft Land and Resource Management Plan*, as amended (Rio Piedras, PR: U.S. Forest Service, 1986).

Dalton, Russell. 1984. *Electoral Change in Advanced Industrial Democracies*. Princeton: Princeton University Press.

Dominguez Cristobal, Carlos. 1989. "La Situación forestal de Puerto Rico durante el siglo XVII." *Acta Científica 3*.

Duncan, Ronald J. 1974. "The Taino as a Symbol of Cultural Identity," *Interrevista* 2:3, (September).

Garcia, Neftalí. 1993. Interviews. New York City (February) and Centro de Información, Investigación, y Educación Social, Hato Rey, PR (December).

Inglehart, Ronald. 1990. *Culture Shift*. Princeton: Princeton University Press.

Jordan, A. Grant. 1981. *Iron Triangles, Woolly Corporatism, and Elastic Nets: Images of the Policy Process*. 1:1, 95-123.

National Archives, San Juan, Puerto Rico. Report of the work done in the survey of the Luquillo National Forest Lands from July 1 to September 30, 1915.

Pico, Fernando. 1986. *Historia General de Puerto Rico*. Rio Piedras, PR: Ediciones Huracán.

Pinto, Benito. 1994. Interview. Former Boating Law Administrator for the Department of Natural Resources, Commonwealth of Puerto Rico, Fajardo, PR (September 10).

Pye, Lucian & Sidney Verba, eds. 1965. *Political Culture and Political Development*. Princeton: Princeton University Press.

Wadsworth, Frank. 1949. *The Development of Forest Land Resources of the Luquillo Mountains, Puerto Rico*. unp. Ph.D. diss. University of Michigan.

Chapter 16

Native Americans and the Desire for Environmental Harmony: Challenging a Stereotype

Isaiah Smithson
Southern Illinois University
Edwardsville, Illinois

The desire for a harmonious relation between humans and the environment is widespread and evidently perennial. The pervasive strength of this fundamental human yearning is exhibited in ancient symbols such as the garden, sacred groves, and the Green Man, and in ancient rituals such as the Eleusinian Mysteries and other vegetation rites. It is also exhibited in modern psychology movements such as ecopsychology and horticultural therapy. Beginning in the 1960s, large numbers of Americans began to conceive of Native Americans as people who, before the arrival of Europeans, lived in harmony with their environment. Unlike the mythic personalities of ancient vegetation rites, Native Americans were conceived as recent, historical examples of humans living at one with nature. In the second half of the 20th century, Native American environmental harmony came to be envisaged as proof for the existence of a lost Edenic past and as a model for a desired paradisal future.

I do not anticipate that 21st-century humans will feel less nostalgia or desire for environmental harmony, nor do I expect 21st-century humans to develop a precise definition of this "harmony" or "oneness" with nature, but I do hope that popular culture and environmental scholarship of the 21st century will move beyond using Amerindians as symbols for this wish. Imagining a vague harmony among Native Americans, the lands they cultivated, the game they hunted, and the natural products they fashioned into artifacts misrepresents Native American history. Maintaining this image also integrates an unreliable level of generalization into environmental education.

Native Americans as the First Ecologists

The early 1960s inaugurated an "environmental decade" in America's history. Books now recognized as landmarks in environmental writing were published, among them Rachel Carson's *Silent Spring* (1962) and Barry Commoner's *The Closing Circle* (1971). The Sierra Club regained prominence by lobbying in court and through the media against commercial and government developments. Membership doubled in the Wilderness Society between 1966 and 1971 and increased sevenfold in the National Audubon Society from 1966 to 1975 (Fox 1981, 315). Environmental legislation began to flow down Capitol Hill: the 1964 Wilderness Act, the 1965 Water Quality Act, the 1967 Air Quality Act, the 1970 Clean Air Act and National Environmental Policy Act, and the 1973 Endangered Species Act composed only part of the legislative river. There were rivers of people, too. Millions of Americans marched down city streets and gathered in parks to celebrate the first Earth Day in April 1970. Environmentalism had become a mass movement.

Native Americans were included in the movement. In *The Quiet Crisis* (1963), then-Secretary of the Interior Stewart Udall wrote about Amerindians' "land wisdom," their "reverence for the life-giving earth," and their being connected with land by "kinship" rather than "ownership" (Udall 1963, 3-5). He observed that the "conservation movement finds itself turning back to ancient Indian land ideas, to the Indian understanding that we are not outside of nature, but of it" (12). In another publication, Udall described Amerindians as "the pioneer ecologists of this country" (qtd. in C. Martin 1978, 160). In 1972, the Advertising Council began a "Keep America Beautiful" campaign consisting of posters and TV spots. The ads reproduced a Native American with a tear trickling down one cheek as he contemplates trash and pollution. According to Iron Eyes Cody (the Cherokee actor featured in the campaign), "The effect it had on the public was amazing, overwhelming. In its second year of worldwide distribution, it tallied up some $30 million in donated air time," surpassing the record of Smokey Bear (1982, 270). Although *Black Elk Speaks* had been published almost unnoticed in 1932, it was reissued in paperback in 1961 to become "one of the most successful books of all time on American Indians" (DeMallie 1984, 58). The book's conception of "two-leggeds" sharing life with "four-leggeds and the wings of the air and all green things" and its insistence that all forms of life are "children of one mother" (Niehardt 1932, 1) were translated into many languages, and the book became "required reading" on many campuses across the United States. Equally popular was a speech attributed to Seattle (Seathl), Chief of the Suquamish and Duwamish. Beginning in the 1970s, versions of this speech were translated into several languages; distributed at environmental rallies and demonstrations throughout the world; reprinted in books, magazines, and newspapers; broadcast over international radio and TV; and included in films (Kaiser 1987, 497-502). It became "probably the single most widely known, respected, duplicated, and repeated statement of the environmental movement" (Rothenberg 1996, 70). Supposedly originating in 1854 and addressed to the "Great Chief in Washington," the powerful speech claims that "Every part of this earth is sacred to my people"; that "All things share the same breath—the beast, the tree, the man"; that "whatever happens to the beasts, soon happens to man"; in effect, that "All things are connected" (Kaiser 1987, 525-30). Although the version of the speech distributed within the environmental movement is now known not to have been delivered by Chief Seattle but to have been written for a 1972 ABC-TV program, the speech eloquently summarizes much popular belief about Native Americans and the environment.[1]

I view this "pioneer ecologists" stereotype as a projection of modern Americans' desire for environmental harmony onto precontact Native Americans. Others have viewed the transformation of Amerindians into ecologists with equal skepticism, doubting that the most recent in a long series of stereotypes is either an authentic rendition or an honorable use of the first Americans:

> Over the past five centuries the American Indian has been called everything from "noble savage" to "besotted alcoholic"....in the heat and froth of the 1960s environmental movement, yet another title—"ecological Indian"—was conferred on the idealized Native American, who was paraded out before an admiring throng and hailed as the high priest of the Ecology Cult. (C. Martin 1978, 13)

Still others have viewed the transformation as enigmatic:

> Scholars have yet to explain the evolution of the popular image of the Indian into that of a conservationist, but clearly, with the arrival of popular environmentalism in the late 1960s and 1970s, Indians had become synonymous for most Whites with conservation. (White 1984, 180)

Generalizing About Native Americans

The stereotype of the precontact Native American as a pioneer ecologist continues into the 1990s, and it is not restricted to popular culture. Current scholarship often makes widely inclusive proclamations about all Amerindian *beliefs* about nature and all Amerindian *uses* of nature. I offer two examples of scholars commenting on religious *beliefs*. Speaking of "all North American Indian groups in the centuries before Columbus"—an exceedingly large number of quite diverse groups—one scholar says, "In spite of the evident differences in tribal ways of life and mythologies, there is an impressive underlying agreement in their expressions of reverence for the earth, kinship with all forms of life, and harmony with nature" (Hughes 1983, 6). Another scholar, evidently referring to the entire millennia-long history of Native Americans, says:

> There are so many Native American religions and their expressions are so varied that it is difficult to generalize upon all Native American religions....Nevertheless, if we keep in mind the distinctive religious differences of individual tribes, some general features are found among most groups. (Hultkrantz 1987, 20)

One of these features is a "similar world view" (20), within which "nature is potentially sacred" (24). I could offer more examples, but most of them would—after cautioning that there are "evident differences" and that it is "difficult to generalize"—nevertheless generalize that all precontact Native Americans revered their environment.

With respect to the second issue, Native American *uses* of nature, current scholarship also offers widely inclusive proclamations; however, generalizations about Amerindian *uses* of nature contradict one another. I offer four examples, the first two disagreeing with the last two as to whether all Native Americans used nature well or abused it:

> The native people were transparent in the landscape, living as natural elements of the ecosphere. Their world...was a world of barely perceptible human disturbance. (Shelter 1991, 226)

> After having studied a mass of evidence in the biological, physical, and social sciences, I am convinced that Indians were indeed conservators. They were America's first ecologists....The Indians [developed] a land

ethic tuned to the carrying capacity of each ecozone. (Jacobs 1980, 49)

By 1492 Indian activity throughout the Americas had modified forest extent and composition, created and expanded grasslands, and rearranged microrelief via countless artificial earthworks. Agricultural fields were common, as were houses and towns and roads and trails [that] had local impacts on soil microclimate, hydrology, and wildlife. (Denevan 1992, 370)

[P]resettlement Indians were not the good conservationists they are reputed to be. Rather than living off the land without depleting it, they altered the landscape profoundly, sometimes beneficially and sometimes not....there is abundant evidence that they overhunted many game species to the point of local extinction. (Chase 1995, 417)

More examples would illustrate the same point. While scholarly generalizations with respect to Amerindian religious *beliefs* about nature converge, generalizations about Amerindian *uses* of nature diverge. I would like to explore these converging and diverging sets of generalizations further by focusing on particular *beliefs* about and *uses* of nature exhibited by particular Amerindian groups.

Sacred Trees, Sacred Crops, Sacred Fire

Most California Native Americans relied on oak trees for food, medicine, and other items (Coulter & Rose 1900, 344; Bean & Saubel 1972, 129-30). As part of their management of oak groves, California Natives engaged in seasonal burning to rid the growing area of diseases borne by rotting leaves, acorns, and other ground matter; to make gathering fallen acorns easier; to clear the oak-growing areas of undergrowth that could fuel a wild fire; and to ensure that cedar, fir, and pines could not compete with the oaks for water, soil nutrients, and light. Most species of mature oaks are relatively fire-resistant. However, competing cedar, fir, and pine seedlings do not rebound readily after repeated burnings (McCarthy 1993, 221). Thus, repeated low-intensity, managed fires create an "optimal environment" for oaks (227), enhancing their distribution, number, and longevity.

Burning was not a secular activity; it was part of a pervasive spiritual relationship that California Natives maintained with the oaks. Many California groups are known to have conceived of trees and other natural phenomena as animate and to have assumed a reciprocal bond with plants and animals. Groups such as the Mono and Chukchansi believed their oak-grove management strategies could be productive only "within positive spiritual relationships between themselves, the land, and the resources it provides" (225). "Plants need people to gather their seeds, and leaves, and roots, and to talk and sing and pray to them....If plants are not used by people, the spiritual relationship is broken, and the plants no longer have a place" (225). In addition to maintaining a daily spiritual relationship with the oaks, the Mono, Chukchansi, Yuki, Miwok, Washo, and other California groups performed specific seasonal ceremonies. Some were performed to guarantee that rain would nourish the trees. Others, termed "first acorn" rites, were performed to give thanks for the new crop of acorns and to seek an "increase of the acorn crop for the following year" (Nomland 1935, 155).

The Mono, Chukchansi, Cahuilla, and other Native Californians venerated and protected oak trees. However, in so doing, they repressed the growth of cedar, fir, and pine trees that could obstruct oak dominance. In effect, these Amerindians used fire to prevent the forest succession pattern that would have evolved had only lightning-induced fires occurred. They created oak orchards at the expense of forest succession. Fire was used not only to enhance acorn yield. According to a study of California Natives, "at least 35 tribes used fire to increase the yield of desired seeds; 33 used fire to drive game; 22 groups used it to stimulate the growth of wild tobacco; while other reasons included making vegetable food available, facilitating the collection of seeds, improving visibility, [and] protection from snakes" (Lewis 1993, 79).

The effects of repeated burning became evident only after the disappearance of California Natives and the institution in the early 20th century of an alternative philosophy of vegetation management—fire suppression. One forest historian commented in 1910 that "with an increasing control of annual fires, the forest and woods...are showing a decidedly aggressive character" (Chase 1995, 223-24). Indeed, another said, "California is wilder now than it was before Europeans arrived. Where there now are forests, there once were vast acorn orchards, painstakingly tended by Native Americans" (G. Martin 1966, 45).

This mixture of *beliefs* expressing reverence for all nature with *uses* supporting only some parts of nature was typical of many Native American groups throughout the continent. Moving from the West Coast to the East

Coast, Eastern Woodland groups replace California Natives, maize replaces acorns as the crop of value, and different types of forests are repressed for the benefit of crops, but the pattern of fundamentally altering a revered environment is repeated.

Maize-Based Agricultural and Ritual Practices

Varieties of Mesoamerican maize (corn) appeared in the Southwest approximately 3,000 years ago. Over several centuries, maize cultivation moved through the Southwest into the lower Mississippi and Ohio valleys, and into other southeastern and northeastern areas with sufficiently long growing seasons. The gradual results were "far-reaching modifications in Indian subsistence patterns and ecological relationships" (Mitchell 1993, 7). Although Amerindian groups continued their seasonal rounds of hunting, gathering, and (in some cases) fishing, tending maize fields caused them to become less nomadic; planting and harvesting rituals developed; and the improved nutrition provided by maize helped Native American populations to expand.

Maize fields were so numerous and so large that they were constantly referred to by explorers, traders, and colonists. Sixteenth and 17th century accounts refer often to the omnipresence of fields, to their extent—sometimes spreading for miles around a village—and to large quantities of stored maize (Maxwell 1910, 79; Day 1953, 330–34; Thomas 1976, 6; Silver 1990, 104; Denevan 1992, 375; Whitney 1994, 104, 118). Many of the open and parklike areas described so often by New World explorers, traders, and settlers were actually abandoned maize fields.

Amerindians girdled countless trees and burned countless forest acres to make way for maize fields (in which beans and squash were normally combined with the maize), and the soil they exposed and cultivated became depleted by this "deep-rooted plant with a high demand for soil nutrients" (Mitchell 1993, 7). Plots were burned and replanted annually for approximately ten years, until the soil could no longer support maize (Thomas 1976, 13). Depleted fields were forsaken and more forest areas were cleared and planted; either the fields were reclaimed after having lain fallow for several years, or they were abandoned altogether. Villages had to be relocated every ten years or so to take advantage of richer soil and new sources of firewood. Abandoned Native American fields—giving way to grass, shrub, and trees—were so numerous that "old fields" became a common term in the colonists' vocabulary.

Like the California Amerindians who integrated a reverent approach into their management of oak groves, most Eastern Woodland groups accompanied their field burning and their planting, tending, and harvesting of maize with veneration and ritual. Planting, rain-making, and harvesting ceremonies were common, as were "Green Corn" celebrations. From the northern boundaries of maize cultivation in New England and New York southward into Florida, the Narraganset, Mohegan, Iroquois, Delaware, Cherokee, Tuskogee, Creek, Osage, Natchez, Seminole, and many other groups celebrated the appearance of the first green corn in late summer (Witthoft 1949, passim). Green Corn ceremonies differed from group to group, but their objectives included spiritual renewal, thanking and propitiating the Corn Mother or other Powers for continued human existence and plentiful plant harvests, conveying hopes for a robust harvest, and expressing hopes for a positive winter.

The results of the maize-based agricultural practices and religious rituals performed in the East were similar to the results of the acorn-based practices and rituals performed in California: preferred plants prospered, other growth suffered, and the ecology was transformed. Recovery from the long hiatus in forest succession caused by maize cultivation is slow and multifaceted. [2] The soil has been burned by fire and blanched by the sun, cultivation has removed organic matter, the maize has depleted essential nutrients, and annual layers of ash have raised the pH level above the acidic range conducive to forest growth (Spurr & Barnes 1980, 220, 286). Therefore, growth required to repair the soil has to initiate the slow succession process. In the southeastern coastal plain, for example, many decades and stages of weeds, grasses, shrubs, and "pioneer" trees (loblolly, longleaf, and slash pines) had to occur before the mixed hardwood and hickory forests that had originally been cleared to make way for maize fields could become dominant again (Williams 1989, 47). It is only in "recent decades" that "pine forests have reached the late stages of their succession, have declined, and have become relatively stable hardwood communities" (48).

In addition to using fire to clear agricultural plots, Native Americans burned forest areas for dozens of purposes—such as the creation of edge habitats for game and as a method of hunting. And Native Americans consumed the forest in fires used for cooking and heating. The results were that Native Americans "destroyed the forest in some places and...modified it over much larger areas" (Day 1953, 343), "selectively altering the species composition of the forest" (Whitney 1994, 17). At best, selective Indian burning promoted the mosaic quality of New England ecosystems, creating forests in many different states of ecological succession, along with ideal habitats for a host of wildlife species (Cronon 1983, 51). At worst, in some areas, "at the period of discovery, the forests had apparently reached the last stage before their fall" (Maxwell 1910, 87). Some areas of the Eastern forests began to return only after European contact; they "recovered and filled in as a result of Indian depopulation, field abandonment, and reduction in burning"; there was evidently "much more 'forest primeval' in 1850 than in 1650" (Denevan 1992, 378, 380).[3]

Conclusions

I offer five concluding thoughts, all grounded in the preceding examples from California and Eastern Woodland environmental *beliefs* and *practices*, and all of them—in my judgment—important to environmental education in the 21st century.

First, the Mono, Cahuilla, Narraganset, Seminole, and other indigenous groups revered all elements of their local environments: the conifers as well as the oaks; the hardwood trees as well as the maize; and fire itself. However, it was not possible for these groups—or any human groups—to have used some parts of nature well without abusing other parts. From the "points of view" of oaks and maize, local Native Americans were conservationists; from the "points of view" of the burned vegetation and suppressed forests, local indigenous people were agents of death. The point seems obvious: when humans support some life forms, they do so to the detriment of others, a point seldom made within environmental education.

Second, if "living in harmony" or "being at one" with nature means developing stories and rituals showing respect for nature, many precontact Amerindian groups lived in harmony with nature. If living in harmony with nature means not altering the landscape, no known Amerindian groups lived at one with nature. There is no evidence to indicate that any Native groups were "pioneer ecologists," if this phrase is meant to imply that they were "transparent in the landscape, living as natural elements of the ecosphere." Indeed, ample "evidence demonstrates that the Indians were not averse to making massive changes in the natural world, as in their burning, when they felt the result would favor their livelihood" (Kupperman 1980, 91).

Third, the fact that no known Native American groups lived without doing extensive damage to forests or some other part of nature does not mean that they damaged their environment to the extent that others coming after them have damaged America's natural resources. Amerindians did destroy forests in favor of orchards, agricultural plots, and game areas. Amerindians used trees for medicines, beverages, food, cooking utensils, housing, canoes, snowshoes, bows, and hundreds of other items. There is abundant evidence that "No single resource was more vital to the native people of North America than trees" (Jonas 1993, 73). Still, indigenous people did not use and abuse forests over several millennia so thoroughly as have the conquering settlers during the last four centuries.

Among the settlers, their increasing population, possession of draft animals, industrial commerce, and Christian beliefs fostered a variety of forest uses and abuses. The "war on woods" referred to by one 19th century immigrant (127) included the export of lumber and naval stores; local manufacture of furniture and other domestic items; charcoal-driven furnaces for innumerable ironworks; thousands of logging sites and saw mills converting trees into lumber for buildings, fences, and roads; steamboats consuming countless acres of trees as fuel; railroads requiring millions of ties and other wood products; and the clearing of forests for cities, towns, farms, and orchards. To deny that Native Americans were "pioneer ecologists" is not to assert that they committed environmental damage on the scale seen in recent centuries.

Fourth, generalizing about precontact Amerindian environmental beliefs and practices is risky because very little is known about many Native American groups who have lived on the continent since the beginning of the current interglacial period; because the physical environments Native Americans lived in changed dramatically over the approximately 15 millennia since the first migrations into what is now the United States; because ecosystems existing at any one time across this huge area differ markedly from one another; and because the United States has a sordid history of negative stereotypes about Amerindians that began in the 16th century and continues into the present.

The Native American-as-ecologist stereotype created in the 1960s does not distinguish between *belief* and *use*; instead, it vaguely implies that Amerindian *respect* for nature entails beneficial *use* of all parts of nature. Extensive evidence indicates that many Native American groups who existed just prior to and at the time of European contact, as well as some other groups existing in earlier eras, experienced nature as animate, conceived of themselves as kin to natural phenomena, perceived the food and raw materials gained from nature as gifts, and responded with religious veneration. However, the examples drawn from the histories of California and Eastern Woodland Native Americans suggest a seeming dissonance between environmental *beliefs* and *practices.*

I see no evidence that precontact Native American *beliefs* committed Native Americans to *practices* consistent with the values of modern environmentalists. Native American *beliefs*, at least in the forest succession examples, seem to have obliged humans, first, to honor nature in general and, second, ritually to acknowledge the *use* or consumption of particular natural entities such as acorns and maize. However, there is no evidence that their religious *beliefs* required them actually to conserve nature or to respect ecological systems—though many Americans have assumed these correlations during and after the 1960s. Although precontact Native Americans had intimate, sophisticated knowledge of their environments, there is no indication that Native American groups felt bound to use this ecological knowledge to preserve the natural communities of which they were a part. "Conservation" is a concept that began to be articulated in America in the second half of the 19th century, while "ecology" became a common term only in the 1960s; there is no indication that traditional Native American respect for nature entailed either of these modern concepts.[4]

Fifth, although Native Americans did not live out the harmonious relation with nature implied by the modern Native American-as-ecologist stereotype, their history does yield three warnings that should become integral to 21st-century environmental education: (1) Native Americans lived with nature for millennia without doing as much environmental damage as modern Americans have effected in just a few centuries. (2) However, their relatively low environmental impact was predicated on a low population base, approximately 3.25 million at the time of contact (Mitchell 1993, 13), and a low level of technology. For a modern world that is both overpopulated and highly technological, these are sobering considerations. (3) Native American history warns that a respectful, even reverential, *attitude* toward nature is not sufficient. Precontact Native American examples illustrate that changes in *behavior* must occur if modern humans are to solve their environmental problems and pursue environmental harmony.

Notes

1. Several discussions of the convoluted history of the Chief Seattle speech are available in popular and scholarly formats. Exposure begins with Kaiser's article (1987) and is continued by Clark (1985), Murray (1993), Rothenberg (1996), and an editorial in *Environmental Ethics* ("The Gospel...." 1989). Kaiser reprints significant versions of the speech.

2. Many abandoned Native American maize fields did not revert to forest, because they were claimed by Europeans as sites for towns and farms. Only in the second half of the 19th century, as industrialization of the Northeast occurred and the Midwest became a locus of agricultural production, did these Eastern pastures and crop lands begin to revert to forest in large quantities (Spurr & Barnes 1980, 444).

3. The extreme reduction of beavers and other fur-bearing animals by Northeastern Native Americans in the 17th century and overhunting of deer by Southeastern groups in the 18th century are other well-known examples of Native American alteration of their environment. However, these instances occur after the introduction of European commerce and during the time in which disease, warfare, and European influences were destroying not only the Amerindian population but also the social and religious fabric of Amerindian groups. However, forest destruction in favor of oak orchards and maize fields begins centuries before the Europeans arrive.

4. For alternative discussions of the seeming contradiction between Ameridian environmental beliefs and practices, see C. Martin's *Keepers of the Game*, Hudson's response to Martin's thesis, Vecsey's article, the two articles by Callicott, the third chapter in Silver's book, and Kay's article.

References

Bean, Lowell & Katherine Saubel. 1972. *Temalpakh: Cahuilla Indian Knowledge and Usage of Plants.* Banning, CA: Malki Museum.

Blackburn, Thomas & Kat Anderson, eds. 1993. *Before the Wilderness: Environmental Management by Native Americans.* Menlo Park, CA: Ballena Press.

Callicott, J. Baird. 1989. *In Defense of the Land Ethic: Essays in Environmen-*

tal Philosophy. Albany, NY: State University of New York Press, 203-19.

_____. 1989. "American Indian Land Wisdom?: Sorting Out the Issues." In Callicott (above), 203-219.

_____. 1989. "Traditional American Indian and Western European Attitudes Toward Nature: An Overview." In *Defense of the Land Ethic: Essays in Environmental Philosophy* (above), In Callicott (above), 177-201.

Chase, Alston. 1995. *In a Dark Wood: The Fight Over Forests and the Rising Tyranny of Ecology.* Boston: Houghton Mifflin.

Clark, Jerry. 1985. "Thus Spoke Chief Seattle: The Story of an Undocumented Speech." Prologue: *Journal of the National Archives 17,* 58-65.

Cody, Iron Eyes. 1982. *Iron Eyes: My Life as a Hollywood Indian. Told to Collin Perry.* New York: Everest House.

Cooke, Jacob, ed. 1993. *Encyclopedia of the North American Colonies.* 3 vols. New York: Scribners.

Coulter, John & J. N. Rose. 1900. Monograph of the North American Unbelliferae. Contributions from the *U.S. National Herbarium 7,* Washington: GPO.

Cronon, William. 1983. *Changes in the Land: Indians, Colonists, and the Ecology of New England.* New York: Hill & Wang.

Day, Gordon. 1953. "The Indian as an Ecological Factor in the Northeastern Forest." *Ecology 34,* 329-46.

DeMallie, Raymond, ed. 1984. *The Sixth Grandfather: Black Elk's Teachings Given to John G. Niehardt.* Lincoln, NE: University of Nebraska Press.

Denevan, William. 1992 "The Pristine Myth: The Landscape of the Americas in 1432." *The Americas Before and After 1492: Current Geographical Research.* Spec. issue of *Annals of the Association of American Geographers 82,* 369-85.

Fox, Stephen. 1981. *The American Conservation Movement: John Muir and His Legacy.* Madison, WI: University of Wisconsin Press.

Hudson, Charles, Jr. 1981. "Why the Southeastern Indians Slaughtered Deer." In Krech (below), 157-76.

Hughes, J. Donald. 1983. *American Indian Ecology.* El Paso, TX: Texas Western Press.

Hultkrantz, Åke. 1987. *Native Religions of North America.* New York: HarperCollins.

Jacobs, Wilbur. 1980. "Indians as Ecologists and other Themes in American Frontier History." In Vecsey & Venables (below), 46-64.

Jonas, Gerald. 1993. *The Living Earth Book of North Americans Trees.* Pleasantville, NY: Reader's Digest Association.

Kaiser, Rudolf. 1987. "Chief Seattle's Speech(es): American Origins and European Reception." In Swann & Krupat (below), 497-536.

Kay, Charles. 1994. "Aboriginal Overkill: The Role of Native Americans in Structuring Western Ecosystems." *Human Nature 5*, 359-98.

Krech, Shepard III, ed. 1981. *Indians, Animals, and the Fur Trade: A Critique of Keepers of the Game.* Athens, GA: University of Georgia Press.

Kupperman, Karen. 1980. *Settling With the Indians: The Meeting of English and Indian Cultures in America, 1580-1640.* Totowa, NJ: Rowman & Littlefield.

Lewis, Henry. 1993. "Patterns of Indian Burning in California: Ecology and Ethnohistory." In Blackburn & Anderson (above), 55-116.

Martin, Calvin. 1978. *Keepers of the Game: Indian-Animal Relationships and the Fur Trade.* Berkeley, CA: University of California Press.

_____. 1981. "The War Between Indians and Animals." In Krech (above), 13-18.

Martin, Glen. 1966. "Keepers of the Oaks." *Discover* (Aug.), 45-50.

Maxwell, Hu. 1910. "The Use and Abuse of Forest by the Virginia Indians." *William and Mary College Quarterly Historical Magazine 19*, 73-103.

McCarthy, Helen. 1993. "Managing Oaks and the Acorn Crop." In Blackburn & Anderson (above), 213-28.

Mitchell, Roberts. 1993. "The Natural Environment." In Cooke (above), 3-14.

Murray, Mary. 1993. "The Little Green Lie: How a Hoax Becomes a Best-selling Book When it Serves a Special Cause." *Reader's Digest* (July), 100-4.

Niehardt, John. 1932. *Black Elk Speaks: Being the Life Story of a Holy Man of the Oglala Sioux.* Intro. Vine Deloria, Jr. Lincoln, NE: University of Nebraska Press.

Nomland, Gladys. 1935. "Sinkyone Notes." *University of California Publications in American Archeology and Ethnology 36*, 149-78.

Rothenberg, David. 1996. "Will the Real Chief Seattle Please Speak Up? An Interview with Ted Perry." *Terra Nova 1*, 68-82.

Shelter, Stanwyn. 1991. "Three Faces of Eden." In Viola & Margolis (below), 225-47.

Silver, Timothy. 1990. *A New Face on the Countryside: Indians, Colonists, and Slaves in South Atlantic Forests, 1500-1800*. New York: Cambridge University Press.

Spurr, Stephen & Burton Barnes. 1980. *Forest Ecology*. 3rd ed. Malabar, FL: Krieger Publishing.

Swagerty, W. R., ed. 1984. *Scholars and the Indian Experience: Critical Reviews of Recent Writing in the Social Sciences*. Bloomington, IN: Indiana University Press,

Swann, Brian & Arnold Krupat, eds. 1987. *Recovering the Word: Essays on Native American Literature*. Berkeley, CA: University of California Press.

"The Gospel of Chief Seattle is a Hoax." 1989. *Environmental Ethics 11*, 195-96.

Thomas, Peter. 1976. "Contrastive Subsistence Strategies and Land Use as Factors for Understanding Indian-White Relations in New England." *Ethnohistory 23*, 1-18

Udall, Stewart. 1963. *The Quiet Crisis*. New York: Holt, Rinehart & Winston.

Vecsey, Christopher & Robert Venables, eds. 1980. *American Indian Environments: Ecological Issues in Native American History*. Syracuse, NY: Syracuse Univeristy Press.

Vecsey, Christopher. "American Indian Environmental Religions." In Vecsey & Venables (above), 1-37.

Viola, Herman & Carolyn Margolis, eds. 1991. *Seeds of Change: A Quincentennial Commemoration*. Washington, DC: Smithsonian Institution Press.

White, Richard. 1984. "Native Americans and the Environment." In Swagerty (above), 179-204.

Whitney, Gordon. 1994. *From Coastal Wilderness to Fruited Plain: A History of Environmental Change in Temperate North America from 1500 to the Present*. New York: Cambridge University Press.

Williams, Michael. 1989. *Americans & Their Forests: A Historical Geography*. New York: Cambridge University Press.

Witthoft, John. 1949. *Green Corn Ceremonialism in the Eastern Woodlands*. Occasional Contributions from the Museum of Anthropology. University of Michigan, #13. Ann Arbor: University of Michigan Press.

Part III

Nature and Culture: Gender and Environment

Chapter 17

The Environment: A Woman's Issue?

Barbara Methfessel
Pädagogische Hochschule
Heidelberg, Germany

The relationship of Woman to Nature and the consequences of that relationship for political and academic policies is a controversial point in the ecological debate. This chapter begins with my view of this debate, then explains *why* and *how* environmental issues are women's issues. I will point to the lack of a strategy of support for this position and the problems resulting from it.

Woman and Nature—A Note on the Ecofeminist Debate

While "concern for environment" in general is a human issue, the ecofeminist debate focuses on the unique relationship of Woman and Nature. There is a general agreement that there is a relationship between women and Nature because both are victims of oppressive control and exploitation by men (Merchant 1980; Fox Keller 1985). There is a nearly general agreement that women are closer to Nature because of their reproductive potential and their involvement with subsistence tasks (I am not here defending an "essentialist" view of Woman). Hotly debated are the questions of (1) whether this "special" relationship is relevant for scientific and political discussions, and (2) whether it exists because of Nature (reproduction), because of culture (socialization), or because of social structures (institutions) (Eichler 1994).

Gender, especially the gender-related division of work and the formation of private and public spheres—which Thompson (1992, 1995) conceptualizes as Hestian and Hermean systems—is perhaps the most relevant factor in the discussion. I agree with Thompson that the issue goes beyond gender (Thompson 1986, 1994). However, it also seems to me that feminist polarization with respect to the influence of "Nature" on men and women leads to a dead end for the following reasons.

• It downplays the influence of culture on Nature. We can't really speak of "Nature" because the nature of humankind is largely a cultural product (for example, the evolutionary changes in our bodies), and it has developed in constant interchange with the "natural world," which is itself a cultural product. Few virgin forests are still "virgin"!

• Pointing out the special relationship of Woman to Nature based on the biological fact of reproduction diminishes Woman's personal and intellectual qualities in comparison with her (admittedly important) role in the reproduction of humankind.

• If we consider the special relationship of women and their interaction with Nature, we must also recognize the other side of the argument, i.e., the special relationship of Man to Nature, a relationship that is also culturally constructed. Such concepts as "man as destroyer" and "woman as carer" lead us into the trap of an ahistorical dualism (Scheich 1994).

• This ahistorical polarization establishes what it criticizes, namely a different relationship of Man and Woman to Nature and a social hierarchy between the sexes. A re-valuing of women's role in reproduction and a devaluing of men's control of Nature do not question the structures that produce that polarization.

• Polarizing men and women ignores the complicity of women in this arrangement and ignores the nature of problems that are part of everyday life.

Concern for the Environment Is a Women's Issue!

Concern for the environment is a women's issue because of (1) pregnancy, childbirth, and motherhood; (2) women's involvement in everyday caretaking; and (3) women's poverty in patriarchal society.

A degraded environment influences human health and the fertility of both men and women. Pollution leaves childless an increasing number of couples who want to become parents. In Germany, this has been estimated at 10-20 percent of all couples. A French investigation showed that the amount of fertile sperm has decreased (*Frankfurter Rundschau* 1995). In Europe, nearly every third pregnancy ends with a miscarriage. Because of German toxic waste, in 1994, no baby was born alive in the Albanian village of Milot (*Frankfurter Rundschau* 1994).

Harmful environmental conditions during fetal development often lead to handicapping conditions.[1] Pollution also increases the number of such environmentally induced or influenced illnesses of children as allergies and cancer. Thirty percent of Germans suffer from allergies. It is estimated that 8 percent (in Germany) and up to 15 percent (in Denmark) of children younger than 15 suffer from forms of eczema, and the tendency is increasing. Some people try to mitigate these facts by pointing out that most allergens are "natural." Is it natural when people react allergically to "natural" food, such as cereals, vegetables, milk, etc.? It would take countless pages to summarize and describe all the illnesses induced by pollution and the residues of herbicides and pesticides.[2]

All these problems are human problems. In actuality, they influence the everyday lives of men directly if they are victims of pollution, and they influence the everyday lives of women indirectly if their husbands, parents, and/or children are victims of pollution. According to Rita Süssmuth, the former German minister of family, about 70 percent of fathers leave their families when a severely handicapped child is born. Women leave their jobs, their outdoor activities, and their leisure activities. All this undermines gender equality.

Households Are Both Victims and Producers of Pollution

Despite the feminist ideal of shared household responsibility, women continue to have a dual responsibility as caretakers and homemakers. It has become their task to protect household and family members from detrimental environmental influences. At a minimum, they must avoid contaminated food and use nonallergenic cleaning materials. They must also act in an ecologically responsible way: buying products that are less damaging to the environment, avoiding and/or disposing of garbage, conserving energy, etc.

For home economists, human ecologists, and environmental educators, it is not enough to advocate an ecologically beneficial household management style or to offer advice to householders. They must also reflect on the political-economic background, the actual conditions, the scope of action, etc., that environmental caretaking requires. This is lacking in most analyses and strategies related directly or indirectly to households and to household management.

In taking responsibility for this area, women are confronted with an increasing amount of household labor. For example, the efforts described above limit the freedom of household decision making and have inevitable consequences for everyday life. The management of a household with members (especially children) who suffer from allergies or chronic diseases demands more resources: more time (not only for everyday care but appointments with physicians, therapy, etc.), more work (a specially adapted lifestyle), more knowledge (of nutrition, toxicology, etc.), more money (medications, special food), and more compensatory management (due to the inability to take advantage of institutions such as nursery schools, fast-food restaurants, etc.).

Since women everywhere are still mainly responsible for household management, this causes conflicts between their private roles as family caretakers and their public roles (paid work, political work, etc.). As women seek to pursue their own interests and are forced by social conditions to take responsibility for their social and economic survival or independence, they can experience irreconcilable conflicts. As stated in the *Beijing Declaration* (1995), "Equal rights, opportunities and access to resources, equal sharing of responsibilities for the family by men and women, and a harmonious partnership between them are critical to their wellbeing and that of their families as well as to the consolidation of democracy" (15).

Problems of the Disadvantaged

Conflicts intensify when disadvantaged people live in the poorer regions of the world. Pollution is everywhere. The degree, as well as the effects and alternatives, differs for the poor. The poor must put up with unhealthy workplaces or living spaces. In German cities, they often live where the environment is most polluted or where the prevailing winds concentrate toxic emissions.[3] There are some regions in East Germany and Eastern Europe that are environmentally (even radioactively) contaminated. People suffer extensively and "die like flies," but they cannot go elsewhere. Poor people have fewer alternatives when deciding on an ecologically beneficial lifestyle. They have less knowledge about how to avoid detrimental conditions or decisions.

Women everywhere belong to the lower socioeconomic strata of society. One-parent families are mostly headed by women. They often live in the worst conditions. Women are the poorest of the poor all over the world. Everyday work is called "women's work," and the care of children is their responsibility. They carry by far the greater burden of these responsibilities. According to the *Beijing Declaration:*

Eradication of poverty based on sustained economic growth, social development, environmental protection and social justice requires the involvement of women in economic and social development and equal opportunities and the full and equal participation of women and men as agents and beneficiaries of people-centered sustainable development. (16)

Precautionary Development: Chance and Risk for the Environment

I do not want to absolve individuals from responsibility. However, Western countries especially must discuss a lifestyle that is based on not only sustainable but also, as German feminists discuss, a precautionary (Jochimsen et al. 1994) development. By this is meant a proactive strategy that would include careful planning within the predictable cycles of Nature (See Schmidt-Waldherr, this volume). This perspective emphasizes a qualitative dimension within the discussions concerning sustainable development (See Piorkowsky, this volume).

Why is caring for the environment a women's issue? Mostly because of the seemingly intractable gender division of labor, which includes a division of responsibility for everyday life. This says nothing about the quality and quantity of women's care and how society rewards this effort. Household work follows (in principle) the rules of sustainable or of precautionary development: orientation to the basic necessities of life and ecological cooperation are principles of household management (Claupein 1994).

Historically and cross-culturally, whether by default (male neglect) or design (male oppression), women have shown a special sensibility to and responsibility for environmental issues. Studies show that girls have a greater awareness of, interest in, and willingness to care for the environment than boys (See McConney & Horton, this volume). Women turn into activists (on a public, as well as on a private, level) in a situation of crisis, as in the fight against the danger of Chernobyl, against pollution caused by local chemical industries, and in the struggle against the sick building syndrome affecting school children (Buchholz 1994). They fight when the health and wellbeing of the family is at stake.

On the other hand, household conditions and household interests too often dictate the horizon of women's activities. I am not here defending those who would contain or confine women to the private sphere. I am speaking not as a theorist but as a feminist. Caring for the environment is objectively in women's essential interests, yet the problems women face in everyday life and the lack of alternatives make them unable to address this concern. Since they obtain no "public" help for "private" problems, women resort to an individualistic way of coping with concrete conditions (Methfessel 1992, 1994). A typical solution in Western countries is to adopt a lifestyle that demands less housework, for example, by taking advantage of fast- and convenience-food or throwaway products, by curtailing family activities, etc. This consumer-oriented (passing as "convenience") lifestyle, in turn, promotes pollution (see Piorkowsky, this volume).

We know this phenomenon also from women of the Third World. Short-term interests of providing everyday necessities dominate long-term interests of protecting the source of these necessities when, for example, they use the wood of the last trees to make fire. The immediate welfare of their own family has to be more important than the long-term welfare of the environment and the community. Their families' survival is more important than an abstract principle. For them, family wellbeing is the measure of good and bad (Wichterich 1994).

In developed societies, status-conferring consumption often dominates household management decisions and is an obstacle to the goal of sustainable development. The old principles of home management (mentioned above) are as unfashionable as "housework" itself. But precautionary household management demands work (mental and physical), i.e., unpaid work in the private sphere that Thompson (1986) calls "Hestian" rather than "female" or "feminine." Devaluation of this domain can be challenged by both feminists and economists:

> Household production is an important part of the output of all nations, yet housework is not recognized when measuring the goods and services that make up the gross domestic product. This undervalues the contribution of women, since they are responsible for most of household production. (Becker 1995, 8)

Responsible decisions about lifestyle, devaluation of consumption, and revaluation of household production and the quality of communities will not be carried out as long as households remain invisible and unaccounted for in strategic proposals, that is, for so long as they are treated simply as a private issue of concern only to women (Claupein 1994).

Conclusions

When they demand change in individual and household behavior, all strategies to protect or save the environment claim the awareness and the responsibility of women. These strategies fail when the necessary preconditions for change are not present.

• It is important to remember that household activities do not alter the causes of destruction. Grading and disposing of garbage, for example, does not solve the problem of wasting natural resources. Effective solutions must begin with production, i.e., with manufacturing and industrial processing.

• Households worldwide must handle the increasing complexity of living conditions (which includes the increasing complexity of both household and emotional work in families). Worldwide, women must cope with work in two areas: the private and the public spheres (Hochschild & Machung 1989; Thompson 1995). Women's resources of time, energy, and income do not increase with the demands of everyday life. Caring for the environment will fail so long as general living conditions and individual efforts exist in opposition.

• Preservation of our natural environment requires new, realistic strategies (Schultz 1994). A successful strategy must include the concerns of households and women. It does not mean that men tell women what to do with respect to household and family matters without reflecting on prevailing living conditions. But both men and women must stop feminizing responsibility in the private sphere.

• It is necessary to recognize and respect the private sphere interests of individuals and households in the public sphere. As long as households remain a "women's" sphere and children are women's chief responsibility, and as long as women serve to compensate for failures of the public sphere, there will be no change. We need a new definition and a new paradigm of private and public (Thompson 1995).

• The ethic of care, empathy, intuition, contemplation, and a special intimacy with living creatures are qualities nurtured in household and family life. They are necessary to promote a caretaking relationship of humankind to Nature (see McNamee, this volume). By taking an active part as family caretakers and "homemakers," both men and women would both acquire these characteristics. As the existing division of labor is a central condition or a basic part of the system of both the exploitation of women and the exploitation of Nature, this also becomes a structural prerequisite.

Individualistic solutions, in our experience, are short-term solutions. They relieve politics from changing oppressive or exploitative structures, and they produce long-term problems. An environment that promotes human wellbeing demands the effort of every individual. Environmental caretaking must become a human issue that involves issues previously labeled solely "women's work."

Notes

1. There are no official statistics. An organization of pediatricians refers to a figure of 10-20 percent of children who are born with greater or lesser handicaps.
2. For a period of three months, I collected relevant articles from a daily German newspaper (*Frankfurter Rundschau*). I got one to three articles per day. Every year the organization of pediatricians claims new measures of toxicity. One must become conscious of the fact that the limit of the "accepted daily intake" is not based on the combination and the accumulation of poisons, and it is calculated for 70-kilo men. Imagine the consequences for children!
3. In Mannheim, for example, a neighboring city of Heidelberg, a city of internationally known chemical industries, such as BASF, the incidence of childhood illnesses caused by pollution (bronchitis, asthma, and other lung diseases) is increasing alarmingly. It follows that illnesses such as scarlet fever can't be "outgrown"; in Heidelberg, too, some children get scarlet fever more than six or seven times (Greiner-Schuster 1994).

References

Becker, Gary S. 1995. "Housework: The Missing Piece of the Economic Pie." *Business Week* (October 16), 8.

Beijing Declaration. 1995. Adopted by the Fourth World Conference on Women: Action for Equality, Development and Peace. Beijing, 15 September 1995. In *Oslo Center Newsletter* No. 7.

Buchen, Judith, et al. 1994. *Das Umweltproblem ist nicht geschlechtsneutral—Feministische Perspektiven.* Bielefeld: Kleine.

Buchholz, Kathrin. 1994. "PCB-Innenraumluftbelastungen in Schulen. Mütter zwischen wissenschaftlicher Risikoabschätzung und Gesundheitsverantwortung für ihre Kinder." In Buchen et al. (above), 168-187.

Busch-Lüty, Christiane, et al. 1994. *Politische Ökologie.* Sonderheft 'Vorsorgendes Wirtschaften'. München: Ökom.

Castleman, Michael. 1996: "Down for the Count." *Mother Jones* (January/February), 20-21.

Claupein, Erika. 1994. "Das Leben anders organisieren. Gemeinsgliche Lebensführung als neue Chance für Männer und Frauen." In Busch-Lüty (above), 60-63.

Eichler, Margrit. 1994. 'Umwelt' als soziologisches Problem. *Das Argument* 205, 359-76.

Greiner-Schuster, Edda. 1994. "Mannheimer Dreck." *Öko-Test 3*, 20-24.

Hochschild, Arlie with Anne Machung. 1989. *The Second Shift: Working Parents and the Revolution at Home.* German edition (1990): *Der 48-Stunden-Tag: Wege aus dem Dilemma berufstätiger Eltern.* Darmstadt: Zsolnay.

Jochimsen, Maren et al. 1994. "Vorsorgendes Wirtschaften. Kontur-enskizze zu Inhalt und Methode einer ökologischen und sozialver-träglichen Ökonomie." In: Busch-Lüty (above), 6-11.

Keller, Evelyn Fox . 1985. *Reflections on Gender and Science.* New Haven-London: Yale University Press. German edition (1986): *Liebe, Macht und Erkenntnis. Männliche oder weibliche Wissenschaft?* München: Carl Hanser.

Merchant, Carolyn. 1980. *The Death of Nature. Women, Ecology and the Scientific Revolution.* German edition (1987): *Der Tod der Natur. Ökologie, Frauen und die neuzeitliche Naturwissenschaft.* München: Beck.

Methfessel, Barbara. 1992. *Hausarbeit zwischen individueller Lebensgestaltung, Norm und Notwendigkeit.* Baltmannsweiler: Schneider.

Methfessel, Barbara. 1994. Arbeit und Zeit im Haushalt als Abstimmungsproblem zwischen Haushaltsmitgliedern und deren Lebenschancen. In Richarz (below), 221-30.

Orland, Barbara & Elvira Scheich, eds. 1995. *Das Geschlecht der Natur.*

Feministische Beiträge zur Geschichte und Theorie der Naturwissenschaften. Frankfurt/M.: Suhrkamp.

Richarz, Irmintraut, ed. 1994. *Haushalten in Geschichte und Gegenwart.* Göttingen: Vandenhoeck & Ruprecht.

Scheich, Elvira. 1994. "Naturbeherrschung und Weiblichkeit. Feministische Kritik der Naturwissenschaften." In Buchen (above), 16-33.

Schultz, Irmgard. 1994. "Das Frauen & Müll-Syndrom-Überlegungen in Richtung einer feministischen Umweltforschung." In Buchen (above), 152-67.

Schulz, Irmgard & Monica Weiland. 1991. *Frauen und Müll. Frauen als handelnde in der kommunalen Abfallwirtschaft.* Sozialökologische Arbeitspapiere 40. Frankfurt/M.: IKO-Verlag für Interkulturelle Kommunikation.

Thompson, Patricia J. 1995. "Reconceptualizing the Private/Public Spheres: A Basis for Home Economics Theory." *Revue Canadienne d'économie familiale 45* :2 (Printemps), 53-57.

_____. 1994. "Dismantling the Master's House: A Hestian/Hermean Deconstruction of Classic Texts." *Hypatia 9* : 4, 38-56.

_____. 1992. *Bringing Feminism Home: Home Economics and the Hestian Connection.* Charlottetown UPEI, Canada. Home Economics Publishing Collective.

_____. 1986. Beyond Gender: Equity Issues in Home Economics Education. *Theory into Practice* (Autumn), 276-83. Rpt. in Lynda Stone, ed. (1994). *The Education Feminism Reader.* Routledge: New York, 184-94.

Wichterich, Christa. 1994. "Pragmatikerinnen des Überlebens. Über den Zusammenhang von Nutzungsrechten und Fürsorgeverant-wortung." In Busch-Lüty (above), 44-46.

Chapter 18

Rural Women and the Ecodynamism of the "Biosphärenreservat Rhön"

Hiltraud Schmidt-Waldherr
Fachhochschule Fulda
Fulda, Germany

The Rhön is a low-mountain region in central Germany, and the "UNESCO-Biosphärenreservat" has been in existence there since 1991. The aim of the "Reservat" is maintenance of the entire biosociological ecosystem of the Rhön (Erdmann & Nauber 1992). Of particular interest are (1) the reciprocal influences of the natural environment and the life circumstances of the residents, and (2) the effect of their activities on nature and the consequences for the environment. The aim of the project is to enable people to live in conscious harmony with nature in their everyday activities. Observation of agricultural methods that have been passed down through history show this goal to be attainable so long as traditional farming practices remain undisturbed. Such agricultural practices not only mirror 100 years' experience of a people's way of relating to Nature; they are also geared toward conservation.

Industrialized agriculture is not feasible in the Rhön because the area has retained its natural life forms, and residents continue to practice traditional farming methods. Also, the customs and sense of identity of a farming community have been preserved—more than in the larger agricultural regions (Brettschneider 1990). Because the Biosphärenreservat project can be adopted only with the concurrence of local inhabitants and not forced upon them, such factors must not only be acknowledged but also reinforced in the practical implementation of the project. In this way, local people can identify with the project and recognize themselves in it. Only then can they take "ownership" of the project.

We conducted an exploratory study on the theme of the Biosphärenreservat, which focused on the circumstances of rural women. There had been no such previous study, so we broke new ground. We employed a qualitative methodology because we wanted rural women to speak in their own voices. In our interviews, as well as in group discussions with them, we asked about their view of Nature, their methods of ecological household management, their subsistence production, and their opinions concerning tourism (Kiunke & Wittmann 1994). We were interested in the following questions:

- How much opportunity did rural women have to apply their socioecological experience and competence in everyday planning and decision making?
- What knowledge of business management do rural women need in order to promote long-term ecologically friendly working and living conditions?
- Surveys of ecological attitudes have always shown conflicts between socioecological awareness and everyday practice. Do rural women also have these conflicts?
- Which needs of rural women in the Biosphärenreservat must be met so that they can assert themselves in actual practice, as well as in the political process?
- How prepared are rural women to put their ideas into action publicly, as well as privately? How ready are they to confront conflict in these two areas?

Subsistence Production and Composting

Rural households in the Rhön are extremely self-sufficient. Raising food is the most important activity. Today, people are buying more and more of their food. Consequently, raising food is decreasing in importance. Field crops and garden crops are both processed and stored. Some women just manage their gardens. Others, living on farms reduced in size, are utilizing what has been left of the farm for their own use. They cultivate vegetables and fruits and keep small domestic animals.

Part of the Rhön is in eastern Germany (the former GDR). Rural women in that region were used to providing fresh fruits, fresh vegetables, and meat for themselves, according to their specific "private home economy." They kept poultry and pigs and cultivated fruits and vegetables which they consumed themselves or sold to state agencies. After unification, "green marketing" became less important (Schambortski 1992). Subsistence production is important now mainly as a way to reduce the cost of living—just as it is in West Germany.

"We are putting very little into the garbage"

Composting and subsistence production are typical in rural households (Claupein 1991). All the women we interviewed operate their farms largely at a subsistence level. The higher the degree of self-sufficiency, the lower the quantity of garbage. By comparison with urban areas, not only is there less garbage in rural areas, but the garbage also contains less packaging material and paper. The women know that packaging creates garbage! But whether or not they try to avoid excess packaging coming into the household in the first place depends on those factors that usually determine consumer behavior (even that of conservation-conscious consumers), i.e., supply and price (Piorkowsky & Rohwer 1988). In some spheres of activity, the women already exhibit a "conservation ethic," e.g., saving water and using detergents known to be safe. A "new" ecological consciousness seems to be developing.

In regions that are shaped by agriculture—such as the Rhön—and are cultivated by small farmers, composting and subsistence production reflect their traditional understanding of natural cycles (Inhetveen & Blasche 1983). Recurrent cycles are taken carefully into account. Seasons—and with them the natural cycles of growth and decline—are observed. Formerly, the food that was the farmers' most important source of sustenance was the result of their own efforts, and it depended on the weather and on the harvest season. Even modern farmers are dependent on the contingency of climate, though no one would starve because of a poor harvest. Nevertheless, the female small farmers we interviewed have maintained their respect for nature, as well as for its unpredictable forces. This makes them careful when dealing with the products of nature, and they cooperate with nature when storing provisions and when composting waste.

Cultivated but Protected, Agricultural Land or Nature Preserve?

There are some regulations proposed by the regional administration that the Rhön women do not understand. One such regulation is the introduction of specified times for mowing meadows. This is to be introduced to a couple of meadows in the Hochrhön—the more mountainous region. The intention is to protect the habitats of rare plants and animals. Theoretically, mowing would be scheduled after the seeds of plants had dropped and the birds had finished breeding.

The growing cycle—which includes mowing times—has been calculated by scientists. The women we interviewed criticized the fact that people living in the area had not been consulted. They argue that local residents should have been asked about how they had cared for the land and the countryside until now. Such knowledge should have influenced planning. The women do not feel their everyday experience has been taken into account. In their view, there is no reason to fix something that is not broken!

"And now just leave it alone…"

According to the Biosphärenreservat concept, some "central areas" in the region would no longer be cultivated. One could not even set foot in them. Nature would be left alone. That is something rural women do not understand. Although they agree that plants and animals need to be protected, they believe the woods, meadows, and slopes should be worked in time-honored fashion. Owners of woods do not understand why they should no longer clear their woods of boughs and scrub (as they always have) to keep down such pests as wood-damaging beetles that would find their habitats improved if they were left alone.

Women's statements make clear that a lot of work must be done by the regional administration to enlighten local inhabitants as to the nature and purpose of such "central areas." By virtue of their traditions and their everyday experience, these women have arrived at a somewhat different understanding of nature. The goal of leaving nature in the "central areas" alone does not fit with their understanding of protecting nature. They are accustomed to protecting the woods and the meadows of the Rhön by taking care of them so as to maintain a natural ecological balance.

Tourism and the Conflict Between Economy and Ecology

To nearby overcrowded cities, such as Frankfurt or Erfurt, the Rhön is an important destination for short holidays. Tourism has become a vital economic factor in the region, yet this activity also brings increasing amounts of traffic. Traffic problems are now being discussed by regional planners. Without a doubt, the huge rush of traffic on weekends is already threatening the success of measures taken to protect the environment. There is not only too much exhaust gas; there are also too many hikers threatening the environment (Planungsbüro Grebe 1993). Protecting rare plants and animals demands a more limited sort of tourism. That is why rural women have proposed enlightening the tourists—for example, by slide presentations—to make them more sensitive to the needs of the natural environment.

"We already have enough of that..."

Basically, there is no way to resolve the conflict between ecology and economy. Planning both to protect nature and to promote tourism requires continuous consideration in regard to the reciprocal effects on both. For example, in the part of the Rhön that belonged to the former GDR, one has to take into account that there is an extremely high rate of unemployment for rural women. Tourism there is still in its infancy. But for many women, tourism makes it possible to rent rooms and earn extra money. None of those women would support serious restrictions on tourism. Until now, these people have not been faced with the disadvantages that come with crowds of tourists.

In the western part, the situation is different. There, tourism is well-developed and established. It is a periodic income source for women, though it increases the amount of housework and keeps women "at home." Most of the "western" rural women do not want any increase in tourism. They feel disturbed by the rush of weekend traffic. They know that the more tourism increases, the more restrictions will become necessary—restrictions that will also apply to them. It will not be easy to balance the number of tourists and the needs of the natural environment (Ott & Gerlinger 1992). It is clear that too many tourists would not only undermine the effort that have been made within the Biosphärenreservat framework but also would ultimately undermine the attraction of the region and thus the basis of tourism itself!

Direct Marketing of Farmer Specialties

Direct marketing helps to revitalize traditional—more ecological—ways of producing food. Such methods of production are more "natural," but they take more time. Baking bread is done according to old recipes, without chemical additives, and jam is made as it was in "Grandma's day" (Pichler 1993). As a result of such old recipes, products have the special flavor appreciated by consumers of farm products. Some of the women suspect that although there is promotion of traditional production, not enough attention will be paid to distribution.

In the western part of the Rhön, the sale of homemade products has been part of everyday farm life. Unlike rural women in the eastern part (who could deliver their products to central collection offices), women in the West have had to sell products on their own. Usually, there was a negative relationship between the work expended and the price received. That explains some resistance to direct marketing. Perhaps such resistance could be overcome if the purchase of some amount of their production at a fair price were guaranteed.

It would be hard to sell homemade products to other local people since other women make their own. So it is necessary to look for sales opportunities outside the region in the big cities. Another advantage is that people there are willing to pay higher prices.

For rural women, the sale of products at green markets, as well as on farms, is time-consuming. Farmers' markets are a tourist attraction. More important (and more cost-effective) would be marketing strategies that could promote the sale of farm women's products. To make direct marketing profitable, traditional—ecological—production must be supplemented by professional marketing.

Direct marketing can be successful only if the conditions of work, as well as the incomes of rural women, improve. In that case, it might become attractive again for young people to take over the parental farm. For now, children are not interested in farming because of low income and lack of leisure time.

Theoretical Concepts Before Practical Experience?

The women's personal experiences with administrative bureaucrats lead them to feel they are not taken seriously. For example, they do not understand why mowing times have to be fixed or why areas have to be closed. Nobody explained their reasons. The administrative agencies just issue regulations; they do not make them understandable, and they do not allow local inhabitants a role in decision making.

The theory of natural cycles—scientific and abstract—is valued above the practical experience of human beings. What for centuries has been maintained without plans and programs is now to be fixed by regulation. These have not been influenced by the local inhabitants' view of nature nor by their everyday knowledge of the country and its plants. For local people to accept programs such as this (or even to identify with them), it would be necessary to give them a substantial role in decision making and in environmental responsibility. The Rhön has become what it is today only because local people have identified with the countryside and have actively worked to shape it.

There is no doubt that the work capabilities of employees in administrative agencies are limited. Nevertheless, it should be the agencies' aim to communicate their goals to the people affected—in this case, rural women. The acceptance of this specific kind of nature reserve depends to a large extent on the flow of information. As far as possible, information should reach people directly from information agencies and should not be distributed hierarchically via "local picks," i.e., mayors or presidents of clubs. As presently constituted, there is the danger of information being filtered or even manipulated.

Until now, there has been no fundamental concept of managing "ecologically friendly" households in the Biosphärenreservat. Such a concept (which could become a point of unification and identification) has yet to be developed. For this, it would be important, above all, to respect such established local customs as subsistence production and composting. From the start, the goal should be to consider both the everyday experiences of the women who are managing their households as well as their lifestyle. Hitherto, a lot of suggestions for household restructuring have been made, especially concerning the handling of garbage and ecologically adequate household management, but none of these suggestions reflect the interests and lifestyles of the women who must do the job (Agrarsoziale Gesellschaft 1990). They are addressed, instead, to a statistical "average woman." In reality, the women enjoy a variety of different living standards. These must be taken into account in the development of ecologically sound concepts, as well as in consultation with those affected by them (Schultz & Weiland 1991).

It should not be forgotten that any kind of ecological household management requires extra time, extra energy, and often extra money. Providing and processing biologically desirable and safe food, and managing waste effectively are large-scale problems. Moreover, ecological housekeeping (as presently understood) is new and innovative, and home managers must, somehow, be kept up-to-date concerning new products, as well as new ways of processing and managing. That again would take extra time and possibly extra patience. Some form of "eco-stress" might result (Thiele-Wittig 1987).

Simply requiring ecologically adequate management will not lead to long-term changes. Any ecological directives not clearly related to the patterns of real thinking and human behavior will not work. People have to be accepted as they are, and their ideas and experiences respected (Schneider et al. 1994). Today, it is especially important to recognize and validate women's practical experience. To create and initiate a framework of conditions that would make ecologically conscious living worthwhile, both the planners and the local population must be cooperatively involved. Only in this way can we realistically expect the "economic" view and the "ecological" view of a "quality of life" to coexist.

References

Agrarsoziale Gesellschaft, ed. 1990. "Organisationsprobleme ländlicher Familien. Ein sozialorganisatorischer Ansatz zur Verbesserung der Lebenssituation von Frauen, speziell der Frauen im ländlichen Raum." AGS-*Materialien Nr. 184*. Göttingen (Germany).

Brettschneider, G. 1990. "Vermittlung ökologischen Wissens im Rahmen des MAB-Programms. In Deutsches Nationalkomitee für das UNESCO-Programm:" *Der Mensch und die Biosphäre*, ed., *MAB-Mitteilungen 32*. Bonn.

Claupein, E. 1991. "Die Lebens und Arbeitssituation von Bäuerinnen. Ergebnisse einer bundesweiten Befragung der Landfrauenverbände im Frühjahr 1988." Münster-Hiltrup (Germany): Landwirtschaftsverlag.

Erdmann, K.-H. & J. Nauber. 1992. "Der deutsche Beitrag zum UNESCO-Programm 'Der Mensch und die Biosphäre.'" In *Deutsches Nationalkomitee für das UNESCO-Programm* "Der Mensch und die Biosphäre," ed., *MAB-Mitteilungen*, 9-57. Bonn.

Inhetveen, H. & M. Blasche. 1983. "Frauen in der kleinbäuerlichen Landwirtschaft: 'Wenn's Weiber gibt, kann's weitergehen...'" Opladen (Germany): Westdeutscher Verlag.

Kiunke, G. & S. Wittmann. 1994. "Land schützen - Frauen nützen." Eine geschlechtsspezifische Studie über Regionalentwicklung im Biosphärenreservat Rhön, dargestellt in den Bereichen Direktvermarktung, Fremdenverkehr und Haushalt: In publication.

Ott, E. & T. Gerlinger. 1992. "Zukunftschancen für eine Region. Alternative Entwicklungsszenarien zum UNESCO-Biosphärenreservat Rhön."

Frankfurt/M. (Germany): Verlag für akademische Schriften.

Pichler, G. 1993. "Lebensmittel aus bäuerlichen Betrieben im Spannungsfeld zwischen Produzenten und Konsumenten." *Förderdienst Produzent-Konsument (Austria), (4, 5, 11).*

Piorkowsky, M.-B. & D. Rohwer. 1988. *Umweltverhalten und Ernährungsverhalten.* Hamburg (Germany): Behr.

Planungsbüro Grebe. 1993. *Rahmenkonzept für das Biosphärenreservat Rhön. Schlußbericht.* Fulda (Germany): unpublished.

Schambortski, H. 1992. "Aber jetzt ist man echt gehändikapt, wenn man eine Frau ist...." Die Vereinigung Deutschlands aus der Sicht von Frauen aus der ehemaligen DDR. Eine Interviewstudie. *Frauenforschung* (Germany), *10* :4, 77-88.

Schneider, L., H. Thomas, W. Hofmann & C. Schneider. 1994. *Zur Ethik des Handelns in Privatwelt und Erwerbswelt am Beispiel von Umwelteinstellungen und Umweltverhalten von Verantwortlichen in Betrieben und privaten Haushalten. Band 2: Empirische Ergebnisse.* Baltmannsweiler (Germany): Schneider Verlag Hohengehren.

Schultz, I. & M. Weiland. 1991. *"Frauen und Müll." Frauen als Handelnde in der kommunalen Abfallwirtschaft.* 2nd ed. Frankfurt/M. (Germany): Verlag für interkulturelle Kommunikation.

Thiele-Wittig, M. 1987. "...der Haushalt ist fast immer betroffen, 'Neue Hausarbeit' als Folge des Wandels der Lebensbedingungen." *Hauswirtsch. Wiss.* (Germany), *35* :3, 119-27.

Chapter 19

Gender Differences in the Environmental Decision-Making of Secondary School Students

Amanda Woods McConney
Andrew McConney
Western Oregon State College
and
Phillip Horton
University of Jos, Nigeria

T his research reflects one subset of a study reported at the 1994 annual meeting of the National Association of Research in Science Teaching. An interdisciplinary curriculum unit was developed that centered on the concept of sustainable development in tropical rainforests. The primary goal of the curriculum was to encourage interdisciplinary thinking in environmental decision-making. The centerpiece of the interdisciplinary unit was the investigation of a simulated environmental problem that required students to develop and then decide on a solution, having weighed a spectrum of possibilities previously explored in class activities and discussions. The results of the study indicate that females were more interdisciplinary in their environmental decision-making when compared to their male counterparts. These results suggest that gender may have been more important than the interdisciplinary approach and content of the curriculum unit. Since gender is not a variable that has been thoroughly examined in environmental research, further study on gender as an attribute in EE research is clearly needed.

Educating for Environmental Decision-Making

Education has not prepared individuals to deal with the complexity and scope of environmental problems (Zoller 1990). In response to this shortfall, there have been continued calls for active learner participation in environmental problem-solving; curricula that are more relevant to current, real-life problems; and EE as a component of both science and liberal education (UNESCO-UNEP 1985; Zoller 1990). Traditionally, the general assumption about environmental education and associated environmental issues was that more knowledge led to an increased awareness of the environment. An increase in awareness was also believed to be associated with a motivation to act (become participants) responsibly toward the environment. Although Hungerford & Volk (1990) acknowledge that in some cases this linear model of environmental awareness leading directly to action may be true, it has not been supported by research in environmental behavior. In general, "issue awareness does not lead to behavior in the environmental dimension" (17). This assertion is especially important in light of the claim that the ultimate goal of EE is responsible environmental behavior (Stapp 1969; Ramsey 1993).

Many instructional materials used in formal educational settings have been designed to provide information and focus only on environmental awareness. Likewise, media efforts in EE have tended to focus on heightened awareness rather than on participation. Consequently, few individuals are exposed to or trained in the skills associated with environmental ownership and empowerment. Hungerford & Volk (1990) therefore advocate the need for an issue-oriented (i.e., thematic, interdisciplinary) model of instruction in EE and suggest that this would facilitate the development of environmentally responsible individuals. In 1969, Clay Schoenfeld (then editor of *The Journal of Environmental Education*) expressed concern over the lack of a structure for EE. Hungerford, Peyton & Wilke (1980) reawakened this concern when they noted that curriculum development in EE was slow to progress from a venture based largely on intuition to one based on research. Almost a decade later, UNESCO-UNEP (1989) repeated the call that should students be instructed beyond ecology basics and environmental awareness to become effective problem-solvers and decision-makers.

Is Gender a Relevant Variable in Environmental Education?

A better understanding of the role of gender in environmental decision-making and problem-solving processes is a concern for environmental educators. In her search for founders of EE, Annette Gough (1994) points out that EE research has been influenced by an Amer-Eurocentric and gendered (largely male) perspective. It is important to understand the transfer of environmental knowledge, skills, and attitudes acquired in the classroom to the learner's decision-making and problem-solving processes. However, just as important is a better understanding of the role gender plays in environmental decision-making and problem-solving processes.

This paper reports on one subset of a study that included the development and implementation of an interdisciplinary curriculum unit centered on sustainable development in tropical rainforests (Woods 1993). With the active participation and advice of teachers, implementation occurred in secondary classrooms to analyze the effects of the unit on the environmental decision-making of secondary school students. The original study was designed to answer three research questions:

1. Will an interdisciplinary tropical rainforest curriculum unit affect the secondary students' approach to solving environmental problems?
2. Does gender play a significant role in the students' approach to solving environmental problems?
3. If gender does influence the students' approach to solving environmental problems, will the influence remain the same after exposure to an interdisciplinary tropical rainforest unit?

This report focuses on the latter two questions regarding gender and solving environmental problems. However, because the answers to the latter two questions are highly related to the answer to the first question, the analysis addresses all three questions.

Design and Procedures: Curriculum Development

The setting for the interdisciplinary curriculum unit developed in this study was constructed from environmental dilemmas that have recently received much public attention—those associated with tropical rainforest loss in developing countries. Fieldwork and extensive firsthand information gathering in the United States, Puerto Rico, and various Latin American countries, such as Costa Rica, Peru, and Belize, allowed the collection of data and real-world examples in the areas of politics and government policies, economics, and tropical rainforest organizations. Along with tropical rainforest ecology, the materials and information gathered during the data collection year were assimilated into appropriate formats for an interdisciplinary curriculum unit targeted at the secondary school level. The components of this curriculum unit included the following topics:

- introduction to tropical rainforest resources: biodiversity
- tropical rainforest resources: economic, social, moral, and esthetic values/ecological services
- introduction to the problem of tropical rainforest loss
- extinction: human population growth and global species; extinction rate
- tropical rainforest ecology
- sources of the problem of tropical rainforest loss: ecological characteristics
- sources of the problem of tropical rainforest loss: economics, agricultural practices, logging practices, social factors, and government policies
- introduction to a tropical rainforest simulated problem
- sustainable development as a balance of alternatives
- What can you do?

The activities in the curriculum unit included a slide/tape presentation, three videotapes, student research and presentations, cooperative group assignments, student discussions, and teacher-facilitated note taking.

Although the objectives of the proposed curriculum unit were not modeled directly on those proposed by Hungerford & Volk (1990), Hungerford, Peyton & Wilke (1980), and UNESCO-UNEP (1989), they reflect similar philosophies. After its initial development, the rainforest curriculum unit was pilot-tested in an introductory environmental science course for nonscience majors ("Survey of Science," SE 1032) at Florida Institute of Technology. The instructor's constructive feedback, along with the researchers' observations of his classes, provided a basis for revisions made in the unit.

Research Design

The study was designed as a quasi-experimental curriculum evaluation (Ary, Jacobs & Razavieh 1985). Nine secondary school teachers (three male, six female) participated in a two-day inservice workshop as training for the implementation of the curriculum. After the training, teachers implemented the interdisciplinary rainforest curriculum unit in the experimental sections of their biology, environmental science, and ecology classes for the prescribed three-week period. The remaining sections of each teacher's classes served as the control group. The random assignment of treatment conditions to the teachers' intact classes resulted in a total of 12 experimental and 12 control sections. The treatment lasted for 15 days (three school weeks, with one day being the equivalent of one 50-minute class).

The comparison curricula were designed by the teachers and consisted of their "normal" lesson plans for ecology. Thus, the comparison curricula were not the same for every teacher, but similar in that ecology was addressed, and social, political, and economic aspects were not included. One researcher conducted observations in control classes to ensure that the comparison curricula were not addressing social, political, and economic issues. Each teacher also completed a daily journal of activities for both their experimental and control classes. Examination of teachers' control group lesson plans, textbooks, and daily journals verified that the comparison curricula did not include the experimental curriculum's interdisciplinary content or approach.

Sample and Instrumentation

Sample

The sample consisted of the 591 students who attended the environmental science, biology, and/or ecology classes taught by the nine high school teachers who participated in the rainforest workshop. The sample comprised 297 females and 294 males. About one third of the students were enrolled in each type of course. The subjects' ages ranged from 13 to 18, with a mean of about 15.5.

Generally, students who take environmental science or ecology classes choose these as science electives; this results in classes with a wide range of student abilities. Students in the environmental science and ecology classes participating in this study were described by their teachers as having average to basic learning abilities. On the other hand, students in 9th-grade honors classes were described as having above-average abilities. Basic biology classes comprised mostly average students. The study sample, therefore, consisted of a diverse group of students described by their teachers as having varying learning abilities.

Instrumentation

Students generally propose unidimensional (simple, possibly reductionist) solutions to complex environmental problems. In order to measure the complexity of students' solutions (by quantifying the degree of interdisciplinarity in their proposed solutions to a defined environmental dilemma), the Approach to Solving Environmental Problems (ASEP) assessment was developed by teachers and researchers working together. The assessment centered on an open-ended essay question that required students to make a decision regarding a contrived but realistic local environmental problem and then to describe the factors that led to that decision.

The ASEP assessment was administered as a pretest to both the control and experimental groups before implementation of the curriculum unit and as a posttest after completion of the curriculum unit. The essays were based on the basis of the number of supported reasons (ASEP supports) that students offered when reaching a decision regarding the environmental dilemma. Each supported reason was awarded 5 points, and no upper limit was imposed on this measure. Essays were also scored according to the number of reason categories (ASEP alternatives) that students utilized in reaching their decisions. Four broad categories were available to students: ecological, economic, moral (including social), and esthetic. Ten points were awarded for each reason category. Therefore, the scoring of ASEP alternatives was done on a closed scale ranging from 0 to 40.

To reduce the risk of experimenter bias, researchers did not score the essay responses. Three independent scorers were hired to read and score the ASEP pre- and post-tests. The scorers received approximately one and one-half hours of training.

To ensure a high degree of agreement among the three scorers and to provide inter-rater reliability estimates, 28 essays were scored by all three raters. The resulting three sets of ASEP scores were correlated in a pairwise manner according to suggestions by Sax (1989, 272). The inter-rater reliability among the three scorers ranged from .70 to .80, with a median value of .80.

Data Collection and Statistical Analysis

Initial data were collected on the students' gender, age, group membership (control or experimental), and pretest ASEP scores. Age and pretest ASEP scores were used as "fairness variables" to level the playing field for all students, while gender and treatment group constituted the main categorical variables of interest for each student. After instruction, data were collected on the number of reasoned supports and the variety of alternatives (interdisciplinarity) found in the students' approaches to solving environmental problems (ASEP-supports and ASEP-alternatives scores).

Results

It was possible (and highly likely) that the amount of supporting evidence (ASEP supports) and the variety of evidence categories (ASEP alternatives) that students used in arriving at and justifying environmental decisions would be correlated. Thus, multiple analysis of covariance (a multivariate statistical procedure that takes this possible correlation into account) was chosen for the examination of this study's results.

This multivariate procedure indicated the significant main effects of both the treatment group and gender on the two aspects (supports and alternatives) of the students' approaches to solving environmental dilemmas. The interaction of the treatment group and gender was nonsignificant. The results further indicated that both age and ASEP pretest scores were useful in leveling the playing field for the purpose of comparing students of differing ages and prior experiences.

These significant results indicated that further exploration of the data was warranted. Follow-up univariate analyses and examination of fairness variable-adjusted means revealed that the treatment-group main effect was carried uniquely by ASEP supports and favored experimental (interdisciplinary) over control (not interdisciplinary) students (experimental mean = 15.0; control mean = 13.0). On the other hand, similar analysis suggested that the gender effect seen at the multivariate level was carried exclusively by ASEP alternatives and favored female students over their male counterparts (female mean = 17.0; male mean = 15.4). When translated to percentiles, these adjusted means indicated that, on the average, students in the experimental group used 11 percent more supports in justifying their environmental decisions than students in the control group. With the same translation, female students typically demonstrated 9 percent more alternatives than males in the categories of reasons (interdisciplinarity) used for justifying their environmental decisions.

In summary, follow-up examination of significant effects showed that the treatment group effect was carried only in ASEP supports, while the gender effect was manifested only in ASEP alternatives. On average, experimental students offered more supporting statements for their environmental decisions as compared to control students. That is, the approach of the interdisciplinary unit was effective in enhancing the number of reasons that students articulated in support of environmental decisions. On the other hand–and this is important for the focus of this study–females (whether in the control or the interdisciplinary group) used more alternative evidence categories than did their male counterparts. That is, regardless of exposure to the interdisciplinary unit, female students demonstrated more interdisciplinarity in approaching environmental problems than did male students.

Conclusions

This study was seen as an investigation of an early step in the development of environmentally responsible individuals, based on the model of instruction in environmental education suggested by Hungerford & Volk (1990) who suggest that an issue-oriented (i.e., thematic, interdisciplinary) model of instruction in EE would facilitate the development of environmentally responsible individuals. However, this research indicates that gender is also an important factor in students' environmental decision-making. The gender effect in favor of females for more alternative evidence categories (ASEP alternatives) is not unique to this study. While investigating Finnish primary school children's preferences in environmental problem solving, Aho, Permikangas & Lyyra (1989) found that girls were able to examine an issue from more points of view than boys. The finding that females use a greater variety of alternative evidence categories in environmental decision-making could imply that interdisciplinary curricular approaches may encourage a greater participation and increased success for females. Gender is not a variable that has been thoroughly examined in EE research. Further research is clearly needed to investigate the relationships between a possible female proclivity for interdisciplinary decision-making and interdisciplinary curricula. Research on gender as an attribute in decision-making is also needed to help guide our EE theory and curriculum development efforts for informal, K-12, and postsecondary programs.

Acknowledgment

Support for the information gathering and initial design of the curriculum unit was provided for the first author by a Pew Fellowship for the Program Initiative: Integrated Approaches to Training in Conservation and Sustainable Development, 1990/1991, Pew Charitable Trusts. The authors also wish to thank Dr. Hilary Swain, Biological Sciences, Florida Institute of Technology, for her review of the curriculum unit.

References

Aho, L., T. Permikangas & S. Lyyra. 1989. "Finnish Primary School Children's Preferences in Environmental Problem Solving." *Science Education 73*: 5, 635-42.

Ary, D., L.C. Jacobs & A. Razavieh. 1985. *Introduction to Research in Education* . 3rd ed. New York: CBS College Publishing.

Gough, A. 1994. "The Founders 'Down Under': Narratives of the environmental Education Movement." Paper presented at the Annual Meeting of the American Educational Research Association, New Orleans, LA. (April).

Hines, J.M., H.R. Hungerford & A.N. Tomera. 1986-1987. "Analysis and Synthesis of Research on Responsible Environmental Behavior: A Meta-analysis." *The Journal of Environmental Education 18*: 2, 1-8.

Hungerford, H.R. & T.L. Volk. 1990. "Changing Learner Behavior Through Environmental Education." *The Journal of Environmental Education 21*: 3, 8-21.

Hungerford, H., R. B. Peyton & R. Wilke. 1980. "Goals for Curriculum Development in Environmental Education." *The Journal of Environmental Education 11*: 3, 42-47.

Ramsey, J. 1993. The Effects of Issue Investigation and Action Training on Eighth-grade Students' Environmental Behavior. *The Journal of Environmental Education 24*: 3, 31-36.

Sax, G. 1989. *Principles of Educational and Psychological Measurement and Evaluation*. 3rd ed. Belmont, CA: Wadsworth.

Schoenfeld, C. 1969. "What's New About Environmental Education?" *The Journal of Environmental Education 1*: 1, 1-4.

Sia, A., H.R. Hungerford & A.N. Tomera. 1985-1986. "Selected Predictors of Responsible Environmental Behavior: An Analysis." *The Journal of Environmental Education 17*: 2, 31-40.

Stapp, W. 1969. "The Concept of Environmental Education." *The Journal of Environmental Education 1*: 3, 31-36.

UNESCO-UNEP 1989. "Developing an Environmental Education Curriculum." *Connect: UNESCO-UNEP Environmental Education Newsletter 14:* 3, 1-2.

UNESCO-UNEP. 1977. "A Comparative Survey of the Incorporation of Environmental Education into School Curricula." UNESCO-UNEP International Environmental Education Programme, *Environmental Educational Series #17*, Federal Republic of Germany.

Woods, A.L. 1993. "Sustainable Development and Tropical Rainforest Loss: The Design and Validation of an Interdisciplinary Environmental Education Curriculum Unit (Dissertation Florida Institute of Technology)." Dissertation Abstracts International.

Zoller, U. 1990. "Environmental Education and the University: The 'Problem-Solving Decision-Making Act' Within a Critical Systems-thinking Framework." *Higher Education in Europe 15*:4, 5-14.

Chapter 20

Ellen Swallow-Richards (1842-1911): Foremother of Environmental Science

Patricia J. Thompson
Lehman College
The City University of New York

E cofeminism derives its theoretical perspective from ecology, notably the principle of interdependence, and it views the common oppression of Woman and Nature, together with concepts of racism and classism, as key concepts. In an article entitled "Feminism and Ecology: Making the Connections," Warren (1987) argues (among other things) that feminist theory must include an ecological perspective and that solutions to ecological problems must include a feminist perspective.

According to Warren & Cheney (1991):

> Ecological feminism is a feminism that attempts to unite the demands of the women's movement with those of the ecological movement in order to bring about a world and worldview that are not based on socioeconomic and conceptual structures of domination. Many ecological feminists have claimed that what is needed is a feminism that is ecological and an ecology that is feminist. (179)

Given these claims, it is surprising that ecofeminists have not familiarized themselves with the life and work of Ellen Swallow-Richards, called by her biographer, Robert C. Clarke, "the woman who founded ecology."[1] In concluding his book, Clarke (1973) observed:

> It is not strange that a woman founded environmental science and formulated a solution to a crisis she saw building, any more than it is strange that a later woman, Rachel Louise Carson, would be credited with reawakening public interest in the crisis long after [Richards] was gone and forgotten .... (255)[2]

Foremothers: A Feminist Issue in the History of Science

The November 1993 issue of *Scientific American* displayed a period photograph of the chemistry faculty of the Massachusetts Institute of Technology (MIT) in 1900: 25 men and 1 woman. The diminutive woman with a clear gaze, hands folded in her lap, sits in the front row—*among* her colleagues but not *of* them. She was Ellen Henrietta Swallow-Richards (1842-1911), the first woman to serve on the faculty of MIT. Ellen Swallow-Richards merits our attention and admiration as a foremother of what would become the environmental movement, even foreshadowing contemporary ecofeminism. In my view, Richards' scientifically grounded environmentalism was the outgrowth of "Yankee frugality"; lessons learned as the classically educated daughter of a small-town farmer and storekeeper; her exposure to astronomer Maria Mitchell during her collegiate years at Vassar; her ongoing research at MIT into air, water, soil, and food; and her work as an industrial chemist and women's educator.

Nature and Necessity

For her first 16 years, Ellen Swallow was educated by her invalid mother and her farmer father. She assumed caretaking and homemaking responsibilities on the family farm in New Hampshire, where, Clarke says, she began her "lifelong pact" with nature (6). At 17, she was sent to Westford Academy, where she received the same classical curriculum—including Greek and Latin—that prepared young men for Harvard College. She was remembered as a bookish girl who, after graduation from the Academy, "hired out" as a teacher, tutor, nurse, and cook and cleaned other people's houses to earn funds for her college education. At Vassar in the 1870s, it was generally believed that education was detrimental to women. Clarke writes, "The 'old maid' or 'spinster' stereotype of an educated woman was widely held. Uneducated men feared her; married women shunned her. Educated men weren't sure" (84). Nevertheless, education—not matrimony—was Ellen Swallow's goal. She wrote her observations of married life to her cousin:

> The silent misery I am discovering. . . among my friends whom I thought happy…makes me shudder. Some things I learned yesterday. . .almost made me vow I would never bind myself with the chains of matrimony…girls don't get behind the scenes as I have, or they could not get up such an enthusiasm for married life. (Clarke 1973, 12)

These insights mark Swallow-Richards' empathy with women's oppressed condition. She saw the need for a wholesome, healthy, life-affirming environment as a woman's right. Her reference to women's "silent misery" and to the "psychosis of negation" (79) are 19th-century precursors of Betty Friedan's "feminine mystique."

Eisenberg (1992) offers an account of the misogynist attitudes that placed serious obstacles in the path of any woman hoping to pursue a career in science:

> Nineteenth-century neuroanatomists and craniologists...diligently measured and weighed female brains to prove women lacked a talent for the hard task of scientific reasoning. Sir David Brewster, Newton's biographer, announced that "the mould in which Providence has cast the female mind does not present to us those rough phases of masculine strength which can sound depths, and grasp syllogisms, and cross-examine Nature." (96)

Women who pursued education were said to contravene the "laws of nature," which prescribed for them the role of mother and housewife. Ellen confronted these attitudes with tact and determination in the course of obtaining an education and pursuing a career in science. She saw nothing incompatible between the activities of the household environment and those of the natural environment. In fact, the two were related systems, and knowledge of this connection, she would insist, should be reliable, based on science, not on superstition.

After college graduation, again working to save money, Ellen Swallow was finally admitted, as a nonpaying "experiment" and "special" student, to the newly established Massachusetts Institute of Technology. Even before she graduated, she had carried out a pioneering project to test the sewage, the streams, and the water supplies of Massachusetts. She had gained an international reputation as a water scientist (Rosser 1992, 42). She received her S.B. in chemistry from MIT in 1873. Although she received an M.S. from Vassar at the same time, opportunities to pursue an advanced degree (Ph.D.) or professional employment in science were virtually nonexistent for women in the 1870s. Consequently, Swallow became an untitled, unpaid, independent scholar and researcher in the laboratories of MIT, where she was able to carve out a niche for women in science.

Despite her earlier renunciation of matrimony, in 1875, Ellen Swallow married Robert Hallowell Richards, a professor of mineralogy at MIT, who had been her teacher. Their household became a meeting place for educated, enlightened women and men. According to Clarke:

> Apart or together, the marriage was a true symbiosis—two lives never intruding but always an advantage to the other. She believed in the synergism of the sexes, that the sum of their relationship should be greater than its parts. This was a physical law. She saw no reason why it should not apply to social nature as well. (56)

The Richards' home reflected environmental awareness in its layout and in the design of ventilation, heating, cooking, and lighting systems—their own Victorian "biosphere." Ellen saw science as a grand opportunity to uplift and upgrade human life through educating women and bringing scientific knowledge into the household.

Bacon's "New Paradigm" for Science

By the time Ellen Swallow graduated from Vassar College in 1870 with a bachelor's degree in chemistry, a "scientific community" was already well established in Europe and America. The foundations of modern science were laid down by the English statesman-philosopher Sir Francis Bacon (1561-1626), who paved the way for the 18th-century Age of Reason and the French Enlightenment. In *Novum Organum* (1620), Bacon had urged scholars to turn from the authority of Aristotle to the evidence of their senses. Bacon addressed his work to the "true *sons* of knowledge," summoning them to woo "female nature's secrets" and to impose male "order and reason" on her "mysteries" (Eisenberg 1992, 96). In *The New Atlantis* (1627), Bacon proposed a scientific workplace ("Salomon's House") with the goal of using tested knowledge for the betterment of human life (Hoobler 1963). In his design for a "normative science," Bacon differed from Descartes (1596-1650), whose method stressed objectivity and reliance on quantification and mathematical operations for the testing of hypotheses. His ideal of a "value-free" science, some feminists now argue, conceals androcentric bias and a "flight from the feminine" (Bordo 1987a, b).

According to Bordo (1987b), the development of Cartesian objectivism (and modern science in general) demonstrates the masculinization of thought, which she describes as the "cultural drama of parturition." Thus, the 17th century nurtured a "scientific revolution" marked by a paradigm shift from the a priori approach of the Scholastics to what we today call "scientific rationalism," embodied in the "scientific method." It was a transition from an organismic to a mechanistic world view, a crucial shift that Merchant (1980) characterizes as the "death of Nature." Keller (1987) acknowledges a "dual theme" pervading the work of all scientists in all ages that is evident in two competing conceptualizations of science as "dominating" and science as "conversing with" Nature (244).

In the latter decades of the 19th century, cities were growing rapidly— and so was pollution. Immigration created crowding and urban blight. Diseases ravaged populations. Opium and cocaine were common ingredients in patent medicines. Large concentrations of populations had historically created unhealthy living conditions. More than human and animal wastes contributed to pollution; industrial and municipal wastes, with their concentrations of dangerous and unfamiliar chemicals, were dumped into the environment:

> There were no laws to slow the tide of filth spewing into streams, rivers, and well waters; rising thick and dark into the air; laying undrained and uncollected in the streets and yards. There was little knowledge of rudimentary sanitation and less concern for its effects. (Clarke, 72)

Basic knowledge of the environment had become an imperative, and environmental study required a different kind of science, a science that connected the natural and the human environments. Ellen Swallow-Richards translated the "sense of order" achieved in a well-managed household (*oikos* in Greek) into an "environmental principle" that could apply to a well-managed planetary home (*oikos*-system/ecosystem) for all humanity. Richards' ecological perspective is an aspect of what I call the Hestian perspective, a perspective that is *oikos*-centered (oriented to the private sphere), as opposed to *polis*-centered (oriented to the public sphere).

By the mid-19th century, Harvard's reputation as a site of liberal arts education was well established, but in an era of industrialization and modernization, there was a demand for Baconian "applied science," generally scorned by practitioners of Descartian "pure science." Harvard's "classical" curriculum was challenged. The creation of a new "institute" in Boston (what is today MIT) presaged the dawn of a new era of technical studies—especially in engineering—in the United States. Harding (1991) states that "the insistence on this separation between the work of pure scientific inquiry and the work of technology and applied science has been long recognized as one important strategy in the attempt of Western elites to avoid taking responsibility for the origins and consequences of the sciences and their technologies or for the interests, desires, and values they promote" (2). This division was as applicable to the scientific community of Swallow-Richards' time as it is today. The difference between "pure" and "applied" science was a belief as strongly held as the difference between women and men. Ellen wanted to change not just the gender makeup of science but the severance of science from everyday life.

Ellen Swallow-Richards saw the problems of the environmental microcosm of the household (the *oikos*-system) as a set of independent yet interrelated problems. In her view, women's ignorance was a major problem to be addressed through education even before suffrage was achieved. In her writing and in her life, we see a dedication to science as a means of empowerment ("increased mastery") for women. While suffragists saw the ballot (exercised in the *polis*-system) as a short cut to political power, Richards thought women should be educated before they voted. The relation of environmental pollution to health had gone unrecognized—at home, in the streets, in businesses, and in industry. According to Rosser (1992):

> Swallow was driven in her work by her outrage over the filth and sewage in the streets and streams of Boston. She studied mineralogy and the chemistry of air to round out her vision of an environmental science encompassing earth, air, and water. (42)

Linking the Human and the Natural Environment

Ellen Swallow-Richards drew on her early farm experiences and her research as an MIT-trained chemist to study the environments in which she believed human beings flourished: the physical, the intellectual, and the moral (Clarke 36). She came to believe that:

> The human-environment interface could be seen most clearly at its source: in the home and in the family. If the human organism is to live in harmony with the environment [Nature], it must be learned at the source. To do that, it is necessary to educate the largest half of the population: Women. (Clarke 79)

By the age of 40, Ellen Swallow-Richards was a world-respected scientist recognized for her studies of samarskite, lead, copper, vanadium, and titanium. In 1879, she was elected to membership in the American Institute of

Mining and Mineralogical Engineers—its first and only woman member (Clarke 98).

Water Study and the Normal Chlorine Map

Water held a central place in Swallow-Richards' concern for environmental quality. It was the first environment of life, the major component of the human body. Working in her Laboratory for Water Analysis—Room 32 at MIT—she launched water studies that continue to challenge environmentalists today. In this, she foreshadowed Rachel Carson's poetic account of "the sea around us." The real story of women and environmental study in the United States began in the 19th century with the work of Ellen Swallow-Richards.

In 1887, the Massachusetts legislature authorized a statewide survey of water and sewage under the direction of recently appointed Dr. T. M. Drown and Richards, who noted wryly that, following the death of her mentor, Dr. W. R. Nichols, she was "still the number two man" (Clarke 144). In this first great scientific survey of pollution, more than 100,000 water samples were analyzed. Richards herself analyzed, in whole or in part, 40,000 of them (Clarke 145). She tabulated her findings on a map of the Great Sanitary Survey. To help her analyze her findings, she "connected the dots" of areas where water chemistry was similar. These lines (isochors) took on a compelling configuration, much like those of today's weather maps. She had created a dramatic image of the state's water pollution. In addition to the inland water-quality markers, new "coastlines" for the state were emerging (Clarke 145-46). When pollution points were connected, they approximated the familiar coastline of Massachusetts. Richards "saw" something in her data. She could extrapolate from it an image of water pollution based on the amount of chlorine present in the water. In present-day usage, we might say that she "conversed" with her subject. From this emergent "picture," it became feasible to determine which impurities in water were natural and which were the products of human and industrial wastes. Today, chlorine is widely recognized as a major source of environmental hazards: furans, dioxins, chlorotoxins, etc. (Sandler 1994, 38).

Swallow-Richards' Normal Chlorine Map is still used to track water pollution. She also created the world's first Water Purity Tables and established the first water-quality standards in the United States (Clarke 147). This led to her work in water and sewage treatment. She thought it "medieval" to discharge raw waste into public waters (152). She published scientific papers on "The Significance of Carbon Dioxide in Potable Waters" and "Permanent Standards of Water Analysis" (189). In 1903, she wrote, "It is hard to find anyplace in the world where the water does not show the effect of human agencies" (190). It would be left to Rachel Carson to bring this fact to public attention.

Ecology: The New Environmental Science of Humanity's Home

In 1873, Ernst Haeckel (1834-1919), a German philosopher and biologist, who had proposed evolutionary theories before Darwin, dubbed a new science oekologie (from the Greek word *oikos*, "household"), which was to be a science of "everyone's home"—humanity's home.

Ellen Swallow-Richards saw that technology was transforming the environment—not always for the better. In 1892, she was invited to speak to the Boot & Shoe Club (made up of the leaders of New England's footwear industry) at the Club's second annual "Ladies' Night." The first year's program was dedicated to the topic "Women in Higher Education." She took the occasion of the second program to propose the introduction of a new environmental science:

> For this knowledge of right living, we have sought a new name ....As theology is the science of religious life, and biology the science of [physical] life...so let Oekology be henceforth the science of [our] normal lives...the worthiest of all applied sciences which teaches the principles on which to found... healthy...and happy life. (Clarke 120)

Without making her views theoretically explicit, Swallow-Richards was promoting a new paradigm for a normative science, a science in which human values were implicit in the course of maintaining a healthful environment. The day after her speech, the *Boston Globe*'s headline proclaimed: "New Science. Mrs. Richards names it Oekology" (Clarke 116). Thus, "pride of place" can be accorded to Ellen Henrietta Swallow-Richards for having introduced the name of the new science of human-environment relations, ecology, to the public over a century ago.

Ecofeminism and Hestian Feminism

Jungian psychologist Ginette Paris links Hestia, the Greek goddess of the hearth, with planetary ecology (1986). To the Greeks, the center of the Earth contained a hestia (hearth), which contained the planet's fire (174). For the geocentric Greeks, the goddess Hestia was associated with the planet Earth, insofar as it is our home, at whose center a fire burns (175).

According to Paris, "In losing geocentrism, we have lost the feeling that this planet is our home," and it now takes ecology, "domestic science," to remind us to take care of our planet (175-176). Christian theology had to dis-place this "earth-bound" spirituality with one that was "heaven-bound." This de-centering was a step toward "mastering" Nature. This dis-place-ment (pushed forward by a science that sends us ever outward and upward) conceals fantasies of escape from "base" necessity, so that "those who are occupied in keeping the hearth fires going, in cleaning, feeding, and taking care of the terrestrial household feel a little neglected" (179). This passage links the "*oikos*-centrism" of Ellen Swallow-Richards, the founder of the discipline of Home Economics (earlier called "domestic science" and in some places today called "human ecology" or "family and consumer sciences"), with the principle of a wholesome planetary environment for the preservation of the human species. The paradigm of Hestian feminism (which I have proposed elsewhere) re-establishes an *oikos*-centric perspective on the systems that make up the human environment (Thompson 1988b, 1988d, 1989, 1994, 1995, 1996).

When Harding (1991) says that women need sciences and technologies that are for women in every class, race, and culture so that they can "learn the existing techniques and skills that will enable [them] to get more control over the conditions of their lives," declaring that the "new sciences are not to be *only* for women," and that it is time to ask "what sciences would look like that were for 'female men'" (5), she is echoing the position that Ellen Swallow-Richards held over a century ago. It is the position I theorize as "Hestian," i.e. *oikos*-centric and humanizing, characterized as empowering, as compared with a *polis*-centric position, characterized as dehumanizing and dominating (Thompson 1994). As Harding (1991) argues:

> Feminism insists that questions be asked of nature, of social relations, and of the sciences [that are] different from those that "prefeminists" have asked, whether conventional or countercultural. How can women manage their lives in the context of sciences and technologies designed and directed by powerful institutions that appear to have few interests in creating social relations beneficial to anyone but those in the dominant groups? (5-6)

These "powerful institutions" (many of which I categorize as "Hermean" in reference to the Greek god of commerce and communications) are those that seek to profit from exploiting natural resources without concern for their effects on our planetary home.

Reflecting on Swallow-Richards

In one of the first books dealing with women's contributions to science, H. J. Mozans (1913) noted that, to Richards, the facts of science were more than uncorrelated facts, and she held this perspective over the 27 years during which she held the post of instructor of sanitary chemistry at M.I.T. Recognized as an authority on air, water, and sewage analysis, Richards used science as a platform from which to promote many programs related to public health. Rosser (1992) summarized her contributions:

> Ellen Swallow Richards....pioneered the testing of air, water, and soil for pollutants. Because of her concern for how the human organism lives in an environment of rapid industrialization, some students of history, notably the environmental engineer H. Patricia Hynes of MIT, consider her the founder of environmental science—as well as a founder of ecology. (43)

In a state-of-the-art review, Harding (1991) argues for the infusion of feminist values and attitudes in science to remove gender and race bias. The selectivity of focus that is based on the assumption that the "male" standpoint is the universal "human" standpoint is now an accepted criticism of male disciplines, including scientific studies. Pushing this argument to its extremes involves paradigm change. In this, Richards was a first. Her "standpoint" was *oikos*-centric, or "Hestian," i.e., ecological. I would call Swallow-Richards the foremother of environmental science and—as such—a Hestian feminist.

Notes

1. Born Ellen Henrietta Swallow in 1842 and married to Robert Hallowell Richards in 1875, she is variously identified as Ellen Swallow, Ellen H. Swallow, Ellen Richards, and Ellen H. Richards. I refer to her here as Ellen Swallow-Richards.

2. According to Clarke, without Ellen Swallow-Richards' work, Rachel

Carson might never have had access to the knowledge she passed on to alert us. Two of the three schools from which Rachel Carson obtained that knowledge had felt the definite influence of the woman who founded environmental science: Johns Hopkins and Woods Hole Marine Laboratory (Clarke 1973, 255).

References

Bordo, Susan R. 1987a. *The Flight to Objectivity: Essays on Cartesianism & Culture.* Albany: SUNY Press.

_____. 1987b. "The Cartesian Masculinization of Thought." In Harding and O'Barr (below), 247-264.

Bubolz, Margaret M., Joanne B. Eicher & M. Suzanne Sontag. 1979. "The Human Ecosystem: A Model." *Journal of Home Economics 71:* 28-31.

Budewig, Carolyn. 1964. "Home Economics in Historical Perspective." Speech before the 35th Annual Meeting of the American Home Economics Association, Detroit, MI, 23 June.

Clarke, Robert. 1973. *Ellen Swallow: The Woman Who Founded Ecology.* Chicago: Follett Publishing Co.

Diamond, Irene & Gloria Feman Orenstein, eds. 1990. *Reweaving the World: The Emergence of Ecofeminism.* San Francisco: Sierra Club.

East, Marjorie. 1980. *Home Economics: Past, Present, and Future.* Boston: Allyn & Bacon.

Eisenberg, Anne. 1992. "Women and the Discourse of Science." *Scientific American 267* (July), 122.

Eckersley, Robyn. 1992. *Environmentalism and Political Theory: Toward an Ecocentric Approach.* Albany: SUNY Press.

Griffin, Susan. 1978. *Woman and Nature: The Roaring Inside Her.* New York: Harper-Colophon.

Harding, Sandra. 1991. *Whose Science? Whose Knowledge? Thinking from Women's Lives.* Ithaca, NY: Cornell University Press.

_____. 1986. *The Science Question in Feminism.* Ithaca: NY:Cornell University Press.

Harding, Sandra & Jean F. O'Barr, eds. 1987. *Sex and Scientific Inquiry.* Chicago: University of Chicago Press.

Harding, Sandra & Merill B. Hintikka. 1983. *Discovering Reality: Feminist Perspectives on Epistemology, Metaphysics, Methodology, and Philosophy of Science.* Synthese Library. Vol. 61. London: D. Reidel.

Holloway, Marguerite. 1993. "A Lab of Her Own." *Scientific American* (November), 94-103.

Hoobler, Icie Macy. 1963. "Is Salomon's House a Modern Utopia?" In Bonnie Rader, ed. 1987. *Significant Writings in Home Economics 1911-1979.* Peoria, IL: Glencoe Publishing Company, 90-5.

Hook, Nancy C. & Beatrice Paolucci. 1970. "The Family as an Ecosystem." *Journal of Home Economics* 62:5, 315-17.

Hunt, Caroline L. 1912. *The Life of Ellen H. Richards.* Washington, D.C.: American Home Economics Association.

Keller, Evelyn Fox, 1987. "The Gender/Science System: or, Is Sex to Gender as Nature is to Science?" *Hypatia* 2:3 (Fall), 37-49.

_____. 1985. *Reflections on Gender and Science.* New Haven: Yale University Press.

McMillan, Carol. 1982. *Women, Reason, and Nature: Some Philosophical Problems with Feminism.* Princeton, NJ: Princeton University Press.

Merchant, Carolyn. 1989. *Ecological Revolutions: Nature, Gender, and Science in New England.* Chapel Hill, NC: University of North Carolina Press.

_____. 1980. *The Death of Nature: Women, Ecology, and the Scientific Revolution.* New York: Harper & Row.

Mozans, H.J. 1913; Rpt. 1974. *Woman in Science.* Intro. by Mildred S. Dresselhauss. Cambridge, MA: MIT Press.

Richards, Ellen. 1904. *The Art of Right Living.* Boston: Whitcomb & Barrow.

_____. 1907. *Sanitation in Daily Life.* Boston: Whitcomb & Barrow.

_____. 1908. *Notes on Industrial Water Analysis: A Survey Course for Engineers.* New York: John Wiley & Sons.

_____. 1910. *Euthenics: The Science of Controllable Environment.* Boston: Whitcomb & Barrow.

Rosser, Sue V. 1992. "The Gender Equation." Rev. of Harriet Zuckerman, Jonathan R. Cole & John T. Bruer, eds. *The Outer Circle: Women in the Scientific Community.* New York: W.W. Norton & Company, and Sandra Harding, *Whose Science? Whose Knowledge? Thinking from Women's Lives.* Ithaca, NY: Cornell University Press. *The Sciences* (September/October), 42-47.

_____. 1991. "Eco-Feminism: Lessons for Feminism from Ecology." *Women's Studies International Forum 14:* 3, 143-51.

_____. 1987. "Feminist Scholarship in the Sciences: Where Are We Now and When Can We Expect a Theoretical Breakthrough?" *Hypatia 2:*3 (Fall).

Sandler, Blair. 1994. "Grow or Die: Marxist Theories of Capitalism and the Environment." *Rethinking Marxism 7:*2, 38-57.

Stone, Lynda, ed. 1994. *The Education Feminism Reader.* New York: Routledge, 184-94.

Thompson, Patricia J. 1994. "Dismantling the Master's House: A Hestian/Hermean Deconstruction & Classic Texts." *Hypatia 9:*4 (Fall), 38-56.

_____. 1995. Reconceptualizing the Private/Public Spheres. *Revue canadienne d'economie familiale 45:*2 (Printemps), 53-7.

_____. 1994 [1986]. "Beyond Gender: Equity Issues in Home Economics Education." In Stone (above), 184-94.

_____. 1989. "Science, Technology and Patriarchal Language." *Feminists in Science and Technology Newsletter. 2:* 2 (February), 1, 8-9.

_____. 1988. "A Hestian Framework for Science and Technology." Paper presented at a Symposium, "Science and Technology: Sexist or Objective?" at the Annual Meeting of the National Women's Studies Association, Minneapolis, MN, June 24.

_____. ed. 1984a. *Knowledge, Technology, and Family Change.* 4th Yearbook of the Teacher Education Section of the American Home Economics Association. Peoria, IL: Bennett & McKnight.

_____. 1988a. *Home Economics and Feminism: The Hestian Synthesis.* Charlottetown, PEI, Canada. The Home Economics Publishing Collective. University of Prince Edward Island.

_____. 1988b. "Hestian Feminism." Paper presented at the Sixth Annual Women's Scholarship Conference: "Women and the Environment: Old Problems, New Solutions." Lehman College, CUNY, The Bronx, NY, March 26.

_____. 1988c. "The Hestian Paradigm: Reconciling Feminist and Family Theory." Paper presented at the Annual Meeting of the National Council on Family Relations, Philadelphia, PA. November 15.

_____. 1988d. "The Nature of Knowledge and the Structure of the University." *The CUNY Women's Coalition Journal 4:*1 (Spring), 1-33.

Tuana, Nancy, ed. 1989. *Feminism &Science.* Bloomington & Indianapolis: Indiana University Press.

Warren, Karen. 1987. "Feminism and Ecology: Making Connections." *Environmental Ethics* 9:1, 3-20.

Warren, Karen J. & Jim Cheney. 1991. "Ecological Feminism and Ecosystem Ecology." *Hypatia* 6:1 (Spring), 179-97.

Williams, Ted. 1987. " 'Silent Spring' Revisited." *Modern Maturity* (October/November), 46-50, 108.

Chapter 21

Wonderment and Wisdom:
The Influence of Rachel Carson (1907-1964)
on Environmental Education

Peter Blaze Corcoran
Bates College
Lewiston, Maine

The words of Herman Melville—"There is one knows not what sweet mystery about this sea, whose gently awful stirrings seem to speak some hidden soul beneath"—have resonated for me in the works of Rachel Carson (1907-1964). In reading her work, I have always felt the presence of some deeper meaning, "some hidden soul beneath." In her descriptions of the sea, we read, at some level, descriptions of life itself. In her warnings of the danger of pesticides to birds, we hear the danger of economic development to life itself.

Carson's humility and tentativeness at what she knew, the way she fully encountered nature, and her respect for the great mystery of the workings of the natural world have encouraged in me a great wondering at what she saw, what she felt—and at her craft in helping us to see it and feel it. Her worry that humankind was destroying the earth and her action to educate and warn others have encouraged in me a deep appreciation for her ecological understanding and for her courage.

In this chapter, I would like to share reflections on Rachel Carson's contributions to environmental education—both her insights into nature awareness and her surpassing grasp of human ecology. I will divide the writing into three chronological periods—the periods of nature writing (1941-1955), wonderment (1956), and environmental problems (1962)—each with its own seminal notions, each as an era in her life, and each as a recapitulation of an era in 20th-century North American EE. By looking at each, I hope to provide some insight into the remarkable woman who educated us through her writing in such a way as to transform the second half of the 20th century and to transform the way we see ourselves in nature, perhaps for all time.

Nature Writing (1941-1955)

Rachel Carson's writing about the sea fulfilled a childhood inland dream. While studying English and science in Pennsylvania, her favorite line of poetry, from Tennyson, "For the mighty wind arises, roaring seaward, and I go…" proved prescient. As she recalled:

> I can still remember my intense emotional response as that line spoke to something within me, seeming to tell me that my own path led to the sea—which then I had never seen—and that my own destiny was somehow linked with the sea. … (Brooks 1972,18)

Captivated by the eternal mysteriousness of the sea, she wrote in such a way as to express the facts, as well as the beauty, of nature—the knowledge, as well as the poetry. According to Paul Brooks (1972), her editor and biographer:

> [S]he was not ashamed of her emotional response to the forces of nature. When a friend confessed to being deeply moved by the "heart-stopping sight" of a flight of wildfowl above the spruces on the Maine coast, she replied: "Don't ever dream I wondered at your tears. I've had the same response too often—perhaps always when alone. The experience I relate in *Under the Sea-Wind* about the young mullet pouring through that tide to race to the sea is one that comes to mind…I didn't tell it as a personal experience, but it was—I stood knee-deep in that racing water and at times could scarcely see those darting, silver bits of life for my tears. Though she had the broad view of the ecologist who studies the infinitely complex web of relationships between living things and their environment, she did not concern herself exclusively with the great impersonal forces of nature. She felt a spiritual, as well as physical, closeness to the individual creatures about whom she wrote, a sense of identification that is an essential element in her literary style. (7-8)

This writing was in the spirit of Anna Botsford Comstock, John Muir, William Burroughs, and others in the nature-study era of North American EE. Through nature-study writing, voice was given both to nature and to the unexpressed sensibilities of readers.

During her years at the U.S. Fish and Wildlife Service, Carson wrote about the sea and began to think of publishing apart from the technical writing of her position. According to Jennifer Logan (1992):

One rejected script became her article "Undersea," printed in the September 1937 *Atlantic Monthly*, and in that one rejection, in a sense, her literary career was born. Her vivid description of the teeming life found in "Undersea" so intrigued one Simon & Schuster editor ....that at his urging Rachel expanded the essay into a book. *Under the Sea-Wind* (1941) was a success with the critics but sold less than 1600 copies in six years....Undaunted, she turned again to the book's central theme: life's eternal cycles as expressed in the eternal sea. For ten years she searched them out, learning scuba diving, taking fishing trawlers to Georges Bank, and earning her daily bread by combining biology with writing as Editor-in-Chief at the United States Fish and Wildlife Service. The result was *The Sea Around Us* (1951), a blend of scientific facts and sweeping poetic prose that spent 86 weeks on the *New York Times* best-seller list, won the National Book Award, and made its modest author suddenly world famous. (2)

The depth and power of her insights and the authority of her research educated readers to a world they did not know. As an educator, Carson shared a subject she believed was vital. Through effective, powerful writing, she vivified the sea for her "students." In the Foreword to the original edition of *Under the Sea-Wind*, she wrote:

> It was written, moreover, out of the deep conviction that the life of the sea is worth knowing. To stand at the edge of the sea, to sense the ebb and the flow of the tides, to feel the breath of a mist moving over a great salt march, to watch the flight of shore birds that have swept up and down the surf lines of the continents for untold thousands of years, to see the running of the old eels and the young shad to the sea, is to have knowledge of things that are as nearly eternal as any earthly life can be. These things were before ever man stood on the shore of the ocean and looked out upon it with wonder; they continue year in, year out, through the centuries and the ages, while man's kingdoms rise and fall. (32)

Carson contextualized her scientific writing within this vastness of time and space. Again and again in her writing we feel the sense of proportion—a soul aware of sitting at the edge of the continent, not only at the edge of the sea but also at the edge of the sea of stars. These vast expanses of time and the cycles of recurring natural events situate her insights of the human place in science. She understood the limits of science—even when she enriched its definition to include humans as feeling and socially responsible participants in its study.

> In accepting the National Book Award for *The Sea Around Us*, she said: The materials of science are the materials of life itself. Science is part of the reality of living; it is the what, the how, and the why of everything in our experience. It is impossible to understand man without understanding his environment and the forces that have molded him physically and mentally. ...The aim of science is to discover and illuminate truth. And that, I take it, is the aim of literature, whether biography or history or fiction; it seems to me, then, that there can be no separate literature of science....The winds, the sea, and the moving tides are what they are. If there is wonder and beauty and majesty in them, science will discover these qualities. If they are not there, science cannot create them. (Brooks 1972, 127-29)

Carson saw the power of science to reveal knowledge of natural processes and to raise questions of the relationship of humans to such processes and to human knowledge of them. Finally, she saw science as needing to evoke the sister of identification and knowledge—personal responsibility.

In addition to scientific knowledge of the "nearly eternal," she would have us feel the poetic essence of our response to nature and of reverence for it. Alas, a scientist who loves nature! This love had been nurtured by two enormously important women in her life—her mother, Maria Carson, and her biology professor and friend, Mary Skinker. Rachel Carson's combined knowledge and love of nature has been compared, in a feminist critique of her writing, to Barbara McClintock's "feeling for the organism" (Hynes 1989, 57).

Another scholar who has plumbed the subtext of Carson's work and its epistemology is Vera L. Norwood (1987). Carson's thinking and feeling lead her to question how we know what we know. She has no godlike perspective apart from nature and human nature; rather, she struggles to locate herself. In an insightful and exceedingly thoughtful piece, "The Nature of Knowing: Rachel Carson and the American Environment," Norwood writes:

> In sum, Carson's work reveals a much more conflicted and complicated approach to nature than her reputation gives her credit for.... Thus Carson alternates between a vision of nature as revered, respected homeplace, to be approached with an almost religious curiosity and as a household existing primarily for production, consumption, coopera-

tion, and management...More often than not, however, Carson is struck by the degree to which the natural world does not function as home *or* household for its human children. Finding herself and her fellows to be outsiders, trespassers in a world that is distinctly "other," she declares both nurturing and managerial responses to nature doomed to miss the point. The occasions when she fails to find herself at home in nature, paradoxically, constitute the high points of her experience. Similarly, the occasions when the economic metaphor shatters against the unwillingness of the natural world to "produce" meaning provide her most telling critiques of human limitations and lead her to doubt all "naming," all artificial boxes into which nature has been "fit." In this context, Carson becomes more than a nature writer; she raises fundamental questions about how human knowledge is constructed, questions that reveal the epistemological hubris underlying much human understanding. These questions prompt her later normative work in *Silent Spring* and *The Sense of Wonder*...For Carson, one of the most important aspects of human interaction with nature is the realization that the protean quality of the natural world cannot be caught by our pattern-hungry minds but that it is our "nature" continually to seek the pattern. In these early books, this fact provides a sort of delicious frustration for her. Aiming to describe both beautifully and accurately the sea- and landscapes, she builds some of her most evocative prose out of foiled attempts at symbolization and categorization. (747-52)

Carson's nature writing deserves to be celebrated for sensitivity, complexity, and depth. She made a significant contribution to EE as the ocean's biographer. She taught us about life in the sea but also standing in reverence of what we don't know. She educated us toward another way to know—that is, to *feel* nature. Finally, she raised questions not only about nature, but also about the nature of knowledge, by which we know nature.

Wonderment (1956)

In 1956, Carson wrote an article for *Woman's Home Companion* entitled "Help Your Child to Wonder." This was the first time in seven years that she had not had a book in production. She wanted to leave the sea for a time; she wrote to her editor, "like that old scorpionlike thing in the Silurian, I have come out on land" (Brooks 201). She had hoped to develop the article as a book, but soon she was to start her research on pesticides, and she never did. The article was published posthumously in 1965 as a book, *The Sense of Wonder.* It is here that Carson is explicitly an environmental educator and here that we can best critique her philosophical and pedagogical contributions to the field.

> What is the value of preserving and strengthening this sense of awe and wonder, this recognition of something beyond the boundaries of human existence? Is the exploration of the natural world just a pleasant way to pass the golden hours of childhood or is there something deeper? I am sure there is something much deeper, something lasting and significant. Those who dwell, as scientists or laymen, among the beauties and mysteries of the earth are never alone or weary of life. Whatever the vexations or concerns of their personal lives, their thoughts can find in the repeated refrains of nature—the assurance that dawn comes after night, and spring after the winter.... (Carson 1956, 88)

In perhaps the most direct statement of Carson's rationale for her kind of EE, she assures us of this deeper meaning, this hidden soul, that lies just beyond our experience in the natural world. So much of education teaches us not to trust our wonder, our intuition, the ineffable sources of our strength. Yet these are part of our knowledge of nature, she says. We hear a validation of the power and authority of childhood experience and an invitation to plumb its depths. And we are enticed to wonderment in the sensual experience of nature. This work, with its explicit inclusion of affect and questions of value, foreshadows the raising of these questions by environmental educators in the 1970s.

> A child's world is fresh and new and beautiful, full of wonder and excitement. It is our misfortune that for most of us that clear-eyed vision, that true instinct for what is beautiful and awe-inspiring, is dimmed and even lost before we reach adulthood. If I had influence with the good fairy who is supposed to preside over the christening of all children I should ask that her gift to each child in the world be a sense of wonder so indestructible that it would last throughout life, as an unfailing antidote against the boredom and disenchantments of later years, the sterile preoccupation with things that are artificial, the alienation from the sources of our strength...If a child is to keep alive his inborn sense of wonder without any such gift from the fairies, he needs the

companionship of at least one adult who can share it, rediscovering with him the joy, excitement and mystery of the world we live in. (42)

I always begin the readings in my undergraduate course in EE with *The Sense of Wonder*. It evokes powerful memories of childhood, and its reading is followed by the writing of an EE autobiography in which the student reflects on the experiences, the teachers, and the places that have provided education about her/his relationship to the natural and human-made environments.

My students keep a journal of reflection throughout the course—of responses to autumnal seasonal changes in nature, to urban environments, and to their own teaching projects. Often, wisdom from Rachel Carson is quoted, followed, and reflected upon in students' journals. Invariably, reading Carson is a significant event that echoes across class discussions and activities throughout the term. I have often wondered why this is so. Why do college women and men find this little volume so powerful?

I believe her encouragement to honor emotions and sensory impressions is compellingly at odds with their college educations. "A feeling of sympathy, pity, admiration, or love..." for the objects of our study is all too rare. Reading her exhortations is provocative as compared to the limits of a college student's own education.

> I sincerely believe that for the child, and for the parent seeking to guide him, it is not half so important to *know* as to *feel*. If facts are the seeds that later produce knowledge and wisdom, then the emotions and the impressions of the senses are the fertile soil in which the seeds must grow. The years of early childhood are the time to prepare the soil. Once the emotions have been aroused—a sense of the beautiful, the excitement of the new and the unknown, a feeling of sympathy, pity, admiration or love—then we wish for knowledge about the object of our emotional response. Once found, it has lasting meaning.

Carson gives us permission to explore the actual and perceived landscapes of childhood again. Young environmental educators do this to make sense of their own childhood wellsprings of wonder. They also can see themselves as adults who will be sources of companionship for a child. A thoughtful consideration of Carson's "sense of wonder" leads students to become present to their remembered landscapes, to their current experience in nature, and to others. Her philosophy of EE reminds us that we are sensory creatures in a sensory world, humble citizens of a mysterious universe, and people free to place ourselves "under the influence of earth, sea, and sky and their amazing life" (1956, 95).

Environmental Problems (1962)

In 1958, Rachel Carson decided to write a brief article on the impact of DDT spraying upon bird life. Her next two and a half years were spent researching and writing one of the most influential books of our time. She told a longtime friend that she had proposed an article about it in 1945. The 1945 article became a book in 1962. In it, she said:

> Only within the moment of time represented by the present century has one species—man—acquired significant power to alter the nature of his world. During the past quarter century this power has not only increased to one of disturbing magnitude but it has changed in charac-ter. The most alarming of all man's assaults upon the environment is the contamination of air, earth, rivers, and sea with dangerous and even lethal materials. This pollution is for the most part irrecoverable; the chain of evil it initiates not only in the world that must support life but in living tissues is for the most part irreversible. In this now universal contamination of the environment, chemicals are the sinister and little-recognized partners of radiation in changing the very nature of the world—the very nature of its life. (1962, 5-6)

Rachel Carson, like Ellen Swallow-Richards before her (see Thompson, this volume), saw the vast destruction of which humans are capable. Ac-cording to H. Patricia Hynes:

> Rachel Carson told students of Scripps College in 1962 that "in the days before Hiroshima," she thought that there were powerful and inviolate realms of nature, like the sea and vast water cycles, which were beyond man's destructive power. "But I was wrong," she continued. "Even these things, that seemed to belong to the eternal verities, are not only threat-ened but have already felt the destroying hand of man." (1989, 181)

Her dedication of *Silent Spring* is instructive. She quoted Albert Schweitzer, "[M]an has lost the capacity to foresee and to forestall. He will end by destroying the earth."[1] She was among the very first to appreciate the gravity of the environmental crisis and her writing in this period precedes all the environmental concern and problem-solving of EE to follow. Destined to be considered a seminal work in environmentalism, perhaps one of the most important books of our time, its writing is meticulously chronicled by Paul Brooks (1972):

> The storm aroused in certain quarters by the publication of *Silent Spring,* the attempts to brand the author a "hysterical woman," cannot be explained simply by the concern of special interest groups for their power or profits. The reasons lie deeper than that. Rachel Carson's detractors were well aware of the real danger to themselves in the stance she had taken. She was not only questioning the indiscriminate use of poisons but declaring the basic responsibility of an industrialized, technological society toward the natural world. This was her heresy. In eloquent and specific terms she set forth the philosophy of life that has given rise to today's environmental movement. (284)

Brooks also quotes a letter to Dorothy Freeman in which Carson captures her feeling of relief upon receiving positive word on the manuscript from a trusted editor at *The New Yorker* ; it was, she wrote, "as if now I knew the book would accomplish what *I long for it to do*" (Brooks 1972, 271).[2]

I think I let you see last summer what my deeper feelings are about this when I said I could never again listen happily to a thrush song if I had not done all I could. And last night the thoughts of all the birds and other creatures and all the loveliness that is in nature came to me with such a surge of deep happiness, that now I had done what I could, I had been able to complete it—now it had its own life..." (271-272).

The years of illness and struggle and research to buttress every argument with fact were at an end. What lay ahead was the vilification of the book by the petrochemical industry and the agricultural establishment. Indeed, the attacks were personal and sought to discredit her as a person, as a woman, and as a scientist. They lasted until her death.

Silent Spring has demonstrated remarkable vitality as a text. It has been translated into nearly every language on the planet. According to Michael Brosnan (Stinnett 1992, 43), 32 years after U.S. publication, it has never been out of print and continues to sell. This book is given credit for changing the way we see our world. According to H. Patricia Hynes (1989), "*Silent Spring* crystallized an 'ethic of the environment' which inspired grassroots environmentalism, the 'deep ecology' movement, and the creation of the Environmental Protection Agency (EPA) and its state counterparts; it influenced the ecofeminist movement and feminist scientists" (9). *Silent Spring* embodies a connectedness with nature, a kinship with other species, and a feeling of the need to take personal action. These are at the heart of EE as it has evolved.

Conclusion

Rachel Carson's three periods of writing, in a sense, recapitulate the brief history of modern EE. The nature-study era, the explicit concern for affect and questions of value, and the problem-solving, action-taking dimension represent the three vital stages of EE.

Carson's writing raises questions about the nature of nature and our knowledge of it; it invites us to stand in wonder at the depth of nature's influence upon our attitudes, and it calls us to responsibility to halt the destruction of nature. Indeed, the intellectual history of EE owes much to the wisdom of this remarkable writer, foremother, and lover of nature.

Notes

1. This is a slight misquotation of Schweitzer, who said, "Modern man no longer knows how to foresee or to forestall. He will end by destroying the earth from which he and other living creatures draw their food. Poor bees, poor birds, poor men...."
2. Recently revealed correspondence between Rachel Carson and Dorothy Freeman reveals the great support and nurturance given to Carson by her relationship with Freeman as she labored over the research and writing of *Silent Spring*.

References

Brooks, Paul. 1972. *The House of Life: Rachel Carson at Work.* Boston: Houghton Mifflin Company.

Carson, Rachel. 1962. *Silent Spring.* Boston: Houghton Mifflin Company.

_____. 1956. *The Sense of Wonder.* New York: Harper & Row.

_____. 1955. *The Edge of the Sea.* Boston: Houghton Mifflin Company.

_____. 1952. *Under the Sea-Wind.* New York: Oxford University Press.

_____. 1951. *The Sea Around Us.* New York: Oxford University Press.

Freeman, Martha. 1995. *Always, Rachel: The Letters of Rachel Carson and Dorothy Freeman 1952-1965.* Boston: Beacon Press.

Greene, Maxine. 1986. College of the Atlantic Commencement Address. Unpublished.

Hines, Bob. 1991. "Remembering Rachel." Yankee (June), 63-66.

Hynes, H. Patricia. 1989. *The Recurring Silent Spring.* New York: Pergamon Press.

Logan, Jennifer. 1992. *An Introduction to Rachel Carson and Her Legacy: A Resource for Maine Educators to Commemorate the 30th Anniversary of the Publication of* Silent Spring. Boothbay, ME: Boothbay Region Land Trust.

_____. 1993. "*Rachel Carson.*" Unpublished manuscript.

Norwood, Vera L. 1987. "The Nature of Knowing: Rachel Carson and the American Environment." *Signs 12,* 740-60.

Stinnett, Caskie. 1992. "The Legacy of Rachel Carson." *Down East* (June), 39-43.

Chapter 22

The Woman/Nature Connection in
Toni Morrison's *Song of Solomon*

Margery Cornwell
College of Staten Island
The City University of New York

Ecofeminist philosophers suggest that there are several interrelated power concepts at work in a patriarchal society. Three of them will inform this work: (1) *Value-hierarchical thinking* gives higher status to what is Up (men) or is gender-identified with what is Up, such as reason, mind, aggression, control, culture, and production, than to what is Down (women) or gender-identified with what is Down, such as emotion, body, passivity, submissiveness, nature, and reproduction. (2) *Value dualisms* organize reality into oppositional and exclusive rather than complementary and inclusive pairs, an "either-or" way of thinking. (3) *Power-over* conceptions maintain relationships of domination and subordination.

Patriarchal frameworks that justify domination of women also justify domination of nonhuman nature by connecting Woman and Nature in terms that "feminize Nature, naturalize women, and position both women and nature as inferior to male gender-identified culture" (Warren 1993, 123). This logic of domination—the Up should obviously have power over the Down—a concept at work in a patriarchal society, is also used to justify domination by race, class, sexual orientation, religion, and age.

Toni Morrison's *Song of Solomon* is a novel about justice and mercy, a mystery tale revealing a variety of responses to the human need to redress grievances, practice forgiveness, and express love. Milkman, an African-American born in Detroit in 1931, moves through the book as the flaccid protagonist, searching for freedom, identity, and gold. I focus here on the significant role that women, primarily Pilate, play in energizing Milkman's search, guiding his initiation, and sharing his discoveries. The women and the men, represented respectively by Pilate and Macon Dead, carry strong symbolic messages in their lifestyles and their means of knowing the world, and these reflect opposing social paradigms. These oppositions are defined in the ecofeminist analysis above. The main body of this chapter examines material from *Song of Solomon* indicating how Morrison emphasizes the dualisms of a patriarchal society in order to illustrate the necessity for their intersection if individual and communal sanity is to be maintained.

The Natural vs. The Material: An Essential Tension

Listen as Milkman's father, Macon Dead, instructs him about his aunt, Pilate, and his future: "I mean for you to stay out of that wine house and as far away from Pilate as you can.... Pilate can't teach you a thing you can use in this world. Maybe the next, but not this one. Let me tell you right now the one important thing you'll ever need to know: Own things. And let the things you own own other things. Then you'll own yourself and other people too" (Morrison 1978, 55).

Macon Dead tries to transmit to his son those traits gender-identified with men, while Pilate presents a set of values often dismissed in a patriarchal society but deemed necessary for health by Morrison and by ecofeminists. The surname of Macon Dead is always used, while the surname of Pilate, which is the same (since she is his sister), is rarely used. Morrison emphasizes Macon Dead's spiritual death because his gaze is forever on the material product; he can only dismiss Pilate's intuitive and earth-directed focus as of no use in this world.

Pilate, who "smelled like a pine forest" (27), looked like a "tall black tree" (38), and sounded like "little round pebbles that bumped up against each other" (40) when she spoke, is, for me and many other readers, the most powerful figure in this book. She wears an earring formed from a small brass box containing the name her illiterate father copied from the Bible because he saw a large figure that "looked like a tree hanging in some princely but protective way over a row of smaller trees" (18). The scent of ginger mysteriously fills the air in her presence, "the air that could have come straight from a marketplace in Accra" (186). This natural fragrance, a sensuous symbol of earth and roots, is also apparent during Milkman's encounter with other women "pilots." Pilate is an African mother whose strength is not power-over, but the power of love, caring, reciprocity, community, and natural simplicity. She "birthed herself" after her mother's death in childbirth; she has no navel and is therefore a natural woman, an Eve.

This physical deformity isolated her as a young woman; she was feared and ostracized. Therefore, she was early forced to "decide how she wanted to live and what was valuable to her. When am I happy and when am I sad and what is the difference? What do I need to know to stay alive? What is true in the world?" (149).

As the French feminists have pointed out, the notion of the completely unitary self is a construct of Western patriarchy. "The integrated self is in fact a phallic self, constructed on the model of the self-contained, powerful phallus. Gloriously autonomous it banishes from itself all conflict, contradiction and ambiguity" (Rigney 1991, 37). Macon Dead may warn Milkman, "You have to be a whole man. And if you want to be a whole man, you have to deal with the whole truth" (70), but Morrison is, of course, parodying him. She knows, and Pilate reflects this knowledge, that this life and identity business is much more amorphous, fragmented, and contradictory than that. Absolutes are deceptive and unreal.

Macon Dead spends his whole life trying to conform to and gain from the white patriarchal society by walking the straight line to calculated success, while Pilate gathers her strength from following her own intuition and personal judgment, which leads her mind through "crooked streets and aimless goat paths" (149). Macon Dead calls her communal home in which she lives with her daughter Rebecca (called Reba) and her granddaughter Hagar "a collection of lunatics who made wine and sang in the streets 'like common street women! Just like common street women!' " (20).

Macon's lonely, obsessive drive toward material wealth drains him of any sense of responsibility for community. "A nigger in business is a terrible thing to see. A terrible, terrible thing to see," says Mrs. Bains (22), one of Macon Dead's soon-to-be-evicted tenants. Macon is terrified at the thought that the white men in the bank—the men who help him buy and mortgage houses—will discover that the raggedy, wine-making Pilate is his sister. He verbally assaults her, "What are you trying to make me look like in this town?" She responds. "I been worried sick about you too, Macon" (20). She absorbs and transforms. While Macon Dead gains his satisfaction from caressing the keys in his pocket, Pilate's major possessions are a bag of bones that hangs from the center of her living-room ceiling, a rock from each of the many states she has visited as a migrant worker, and a geography book.

To Macon Dead's disgust, Pilate, her daughter, and her granddaughter live in Detroit with light from candles and kerosene and heat from wood and coal, pumping water into a dry sink through a pipeline from a well and living, in his words, "pretty much as though progress was a word that meant walking a little farther on down the road" (27). For him, progress (as defined by the dominant society whose goals he embraces) is symbolized by acquisition of material wealth.

Twelve-year-old Milkman knows that neither the supposed wisdom of his father nor the taunts of his schoolmates can keep him from this woman Pilate in black, with a brass box dangling from her ear, who serves his friend, Guitar, and him a perfect egg when they dare to venture into her house, a place of peace, energy, and singing. "...It was the first time in his life that he remembered being completely happy....No wonder his father was afraid of them" (43). And his father is not always able to completely reject her power either as he stands in the dark gazing through the window into a house he will never enter. "He wanted no conversation, no witness, only to listen and perhaps to see…the source of that music that made him think of fields and wild turkey and calico" (29). This is a wistful acknowledgment of paradise lost, the price paid for complete acceptance of material gain.

Macon remembers their lives as children on the farm of his father, the first Macon Dead, a freed slave. His father had made a paradise out of the land he purchased piece by piece in Danville, Pennsylvania, the land he called Lincoln's Heaven. The son had worked side by side with his father, for whom he was named, until the day a white neighbor shot his father from behind in order to obtain his land. Macon was sixteen; Pilate was twelve.

Morrison emphasizes the misguided direction Macon Dead takes in order to avenge the death of his father. He seeks to show the white establishment that he can be good at their game of acquiring, owning, controlling. He pays "homage to his own father's life and death by loving what his father had loved: property, good solid property, the bountifulness of life" (304). But he loves owning to excess and forgets the life. "Grab it. Grab this land! Take it, hold it...make it...shake it, squeeze it, turn it, twist it, beat it, kick it, kiss it, whip it, stomp it, dig it? plow it, seed it, reap it, rent it, buy it, sell it, own it, build it, multiply it, and pass it on" (237-238). In this attitude rest the seeds of the violence of ownership, of domination of people and land. When Macon Dead and Pilate leave the land, Pilate retains the sense of love and mutuality that her father felt toward it; Macon retains only the pride of ownership.

Macon Dead's death of spirit sifts down on the women in his life as a grown man, his wife, Ruth Foster Dead (whom he married because she was the daughter of a doctor), and his two daughters, Magdalene, called Lena, and First Corinthians. "Under the frozen heat of his glance they tripped over doorsills and dropped the salt cellar into the yolks of their poached eggs" (10). He has deprived his wife of any sexual attention for decades because he suspects there was something "unnatural" between her and her father, whom she worshipped. Provided with an aphrodisiac by the life-supporting Pilate, who sees Ruth "dying of lovelessness" (157), Ruth tempts Macon briefly to conceive Milkman. However, in his shadow Ruth cannot provide suitable nourishment. "She did not try to make her meals nauseating, she simply didn't know how not to" (11). Lena and Corinthians, grown women, well educated and paraded through their lives like the "virgins of Babylon" (218) by their father, spend their sterile days in their home, "more prison than palace" (9), making bright, lifeless flowers out of velvet.

Pilate guides Milkman away from this doomed house, where he hears only criticism and commands and practices only self-indulgence and indifference. He follows a trail from Michigan to Pennsylvania to Virginia, first in search of gold and finally in search of family and community. In Shalimar, Virginia, he sees women who don't carry purses and look like what Pilate must have looked like as a girl. Milkman's city clothes are gradually destroyed, and he loses his watch; he begins to investigate a new kind of wisdom, the wisdom advocated by his aunt in opposition to that of his father, whose attitude toward Nature and Earth is that of an entrepreneur, not a sharing participant. He hears the local men communicating in the dark woods in a language of "a time when men and animals did talk to one another....And if they could talk to animals, and the animals could talk to them, what didn't they know about human beings? Or the earth itself, for that matter" (281).

In this Southern land, Milkman also learns about the reciprocity of love. Previously, he has selfishly loved and abandoned Hagar, Pilate's granddaughter. She dies after a frenzied shopping spree, buying items she thinks will make her more attractive to him. Ironically, her rush to achieve some commercial concept of beauty occurs at the same time that he is being initiated into a simple, shared kind of love, the kind of love that compels him to wash dishes for the first time in his life.

Pilate's means of redressing grievances is not death-producing like that of Macon Dead or Guitar, Milkman's friend, who is a member of a secret organization of black men who kill whites in revenge for black deaths. Neither is hers passive. Rather, it is educational, embracing life and love. After she hits Milkman over the head and throws him in the basement for causing the death of her beloved granddaughter, she forgives him and gives him a box of Hagar's hair and some instruction: "You can't take a life and walk off and leave it. Life is life. Precious. And the dead you kill is yours" (210). The Biblical *Song of Solomon* chants, "Love is as strong as Death." Stronger, asserts Morrison. After Pilate forgives, she goes with Milkman to bury the bones of her father, inadvertently saving Milkman's life by giving hers when Guitar tries to shoot him over a misunderstanding about gold.

A Message of Reciprocity

Some critics have suggested that Pilate is not a powerful figure but a foolish or unconscious one (Davis 1990, 21; Bakerman 1981, 563). After all, it takes Milkman's synthesis of information to identify the bones in her bag and to discover that her ghost father's command to her, "Sing," is a reference to her Native American mother's name, Singing Bird, rather than an imperative to act. However, in her confusion resides the message of reciprocity that the book promotes. Without her intuition and sensitivity, Milkman would not have amended his search from material gold to historical and communal "gold." Conversely, without his knowledge she could not have properly identified her father's bones. Even though she had misunderstood her father's instruction, she has made a gift of her understanding of his command, "Sing."

Loving and flying, skillfully executed, seem natural, effortless acts. However, each takes training and nurturing. Love and flight are important in *Song of Solomon*, and they are often connected—in negative and positive ways. Milkman always yearns to fly physically and must learn from Pilate how to fly spiritually. The novel opens and closes with both loving and flying. At the beginning a man "flies" off a building, leaving a note saying, "I loved you all" (3). This alarms Milkman's pregnant mother, hastening his birth. At the conclusion of the book, as Pilate is dying, she says, "I wish I'd a knowed more people. I would of loved em all. If I'd a knowed more, I would a loved more" (340). This is her ultimate message of inclusion. Milkman realizes that she can fly without leaving the ground, and understanding her message, he is able to fly. He knows his father will never be interested in the "flying part," although part of his education has been to gain some understanding and tolerance of his father's goals while not accepting them for himself. Some of Pilate's earlier instruction to Milkman involves the need to embrace contradictions in loving and flying. "You can't fly away and leave a body" (336). The actual song of Solomon, which Pilate sings throughout the book and the children of Shalimar use as a playgame, emphasizes the problem: "Solomon don't leave me here/Cotton balls to choke me/Solomon don't leave me here/Buckra's arms to yoke me. Solomon done fly, Solomon done gone/Solomon cut across the sky, Solomon gone home" (307). Solomon, Milkman's great-grandfather, was a flying African who flew away from slavery back to Africa and freedom. He abandoned his wife and 21 children. She ultimately went insane. Flying can be a release, expanding physical and spiritual horizons, but if one flies or lives only for individual freedom and gain (as do Solomon and Macon Dead), forgetting the community, a violation has occurred. Freedom must be connected with responsibility. The same is true of those who love too much, who possess or who lose their own identity in love, who see themselves as commodities. They often die or go insane. They are "lifted" in love. Theirs is an extreme, lacking balance, focusing on one individual's need alone.

Conclusions

Toni Morrison has said that she chose a male protagonist in *Song of Solomon* because she thought he had more to learn and that she was amazed at how much the women had to teach him (McKay 1983, 410). Macon Dead and Milkman, before his education by Pilate, represent a society that embraces domination in the name of linear material progress, a society that goes for the gold and values that which is gender-identified with men. It is in need of the tempering influence of the elements gender-identified with women, which the dominant society suspects and often dismisses, elements essential and powerful. Throughout *Song of Solomon*, Pilate provides sustenance for Milkman in opposition to the wish of Macon Dead and Guitar for his destruction; she makes his conception possible, thwarts Macon's attempts at abortion, foresees the date of his birth, predicts that a woman will save his life, and becomes that very woman who receives Guitar's gunshot, meant for Milkman. She officiates over his physical and spiritual life, saving him from the aggression, control, and greed of one man who resents his life from its inception and another who warns, "Your day has come."

At the end of the novel, when Milkman leaps toward Guitar from his cliff in the dark, "as fleet and bright as a lodestar," he knows these two things: (1) "If you surrendered to the air, you could ride it." If you do not try to control it, but rather to understand and work with it, you can fly. (2) "Here is my life. Take it if you need it. I do not fly to escape but to serve" (341).

Pilate had shown him the sky, "the blue of it, which was like her mother's ribbons, so that from then on when he looked at it, it had no distance, no remoteness, but was intimate, familiar, like a room that he lived in, a place where he belonged" (211).

The concluding sentence of *Song of Solomon* carries us back to the introductory comments of this paper. If we humans are to maintain individual and community sanity, we must make the dualisms of our contemporary society, as represented by Pilate and Macon Dead, intercept and complement. While the real strength in this book resides with Pilate, she uses it not as the "power-over" of a patriarchal society, but as a force to provide instruction and love. The patriarchal society, whose goals Macon Dead embraces, advocates power as domination in the name of linear material progress, valuing that which is gender-identified with men. This dominant society needs the tempering influence of the elements gender-identified with Woman and Nature while it often suspects and dismisses them. It must learn, as does Milkman through Pilate's instruction, that the land on which we place our feet must not simply be owned and plowed but must be respected and protected.

References

Bakerman, Jane S. 1981. "Failures of Love: Female Initiation in the Novels of Toni Morrison." *American Literature 52*:198, 541-63.

Davis, Cynthia A. 1990 "Self, Society, and Myth in Toni Morrison's Fiction." In Harold Bloom, ed. *Modern Critical Views: Toni Morrison,* 7-26. New York: Chelsea House.

McKay, Nellie. 1983. "An Interview with Toni Morrison." *Contemporary Literature 24* (Winter): 413-29

Moi, Toril. 1985. *Sexual/Textual Politics: Feminist Literary Theory.* New York: Methuen.

Morrison, Toni. 1978. *Song of Solomon.* New York: Signet.

Rigney, Barbara Hill. 1991. *The Voices of Toni Morrison.* Columbus: Ohio State University Press.

Warren, Karen J. 1993. "A Feminist Philosophical Perspective on Ecofeminist Spiritualities." In Carol Adams, ed. *Ecofeminism and the Sacred.* New York: Continuum, 119-33.

Chapter 23

Children's Ideas of Nature:
The Social Construction of a "Nature Set"

Luigina Mortari
Istituto di Scienze dell'Educazione
Verona, Italy

The cognitive representations that human beings have about the natural world considerably condition their relationship with the elements of Nature. These representations imply criteria of judgment that are capable of authorizing or limiting determined behavior. Modern culture is fed by a Cartesian concept of Nature: an inert and passive thing, uniform and mechanical, broken into separate entities, detached from human beings and inferior to them. To think of Nature as a lifeless world, distinct from human affairs, like a storehouse of resources that can be endlessly exploited, is different from thinking of Nature as a vital world. The birth of a new ecological culture presupposes a new idea of Nature as a dynamic and "autopoietic" system in which everything is interconnected, a vital breath that passes through all things, including the human beings who must conceive of themselves as beings indissolubly bound to the Earth.

"To be part of nature is already a wonderful thing" (Elisa, 10)

The ecological crisis requires a radical change in thought. Therefore, environmental education must above all encourage new ways of thinking about Nature and the place of human beings in the natural order. To do so, it is necessary to examine the ideas, deeply rooted in our culture, that condition the dominant worldview. These ideas constitute the underlying structures of our thinking and thereby define the horizon of intelligibility of our experience. They generate our attitudes and direct our actions. Therefore, it is important for those who work in EE to know the "maps of ideas" or "cognitive maps" with which children relate to Nature. The aim of this exploratory study was to answer two questions:

- Does the child's concept of Nature involve the vegetable and animal kingdoms only, or does it include humans?
- How does the child evaluate human behavior toward Nature?

Sample and Method

The research subjects live in a rural area of northern Italy where intensive farming has brought about considerable change in the landscape. Villages are surrounded by cultivated fields, and there is no natural woodland. The sample were primary school children aged 6 through 10 attending 1st, 2nd, and 4th grade respectively. The 208 subjects were divided into 11 groups (two 1st grades, four 2nd grades, and five 4th grades). I met each group twice in a whole-class setting.

During the first meeting, the children were asked to "draw Nature" and to accompany the drawing with a comment. The second meeting, which took place several days after the drawing, aimed to stimulate a discussion about the meaning of the word *Nature*. The children were asked to describe Nature to a fictitious boy and girl from a distant planet. The children were invited to put on the blackboard the Nature set taking shape through the interchange of different interpretations.[1] The older children were asked to compare the preliminary drawings with the results of the discussion to ascertain what differences there were and to understand the reasons for them. Discussions were taped and transcribed.

Analysis of the Drawings and Discussion

What elements of the children's drawings identify Nature? From an analysis of the preliminary drawings and of the brief comments accompanying them, it seems that children of different ages share a similar idea of Nature. In general, they tend to identify Nature with the plant world (100 percent of the drawings show this) and with the animal world (around 70 percent of the drawings include some animals). In a few drawings, especially those of 6-, 7-, and 8-year-olds, a house appears. Not one drawing by a 6- or 7-year-old shows a human being, whereas 12 percent of the older children drew human beings. This idea of Nature as a world of plants and animals is confirmed also by the answers given by the children at the beginning of the discussion, when they were asked to describe their drawings:

Davide (6): Nature is made of trees, animals, flowers, and grass.

Stefania (7): It's made of plants, birds, butterflies.

Marco (7): of water, grass, flowers, plants and animals.

Valeria (6): It's made of trees, flowers, animals, and leaves.

Only one child held an exotic image of Nature:

Matteo (7): It's made of lions, elephants, and tigers.

The fact that, in some of the older children's drawings, human beings were represented might lead to conjecture that a different idea of Nature would emerge from the conversations. Instead, we find that the image of the natural world as expressed by the younger children is confirmed: Nature is made of plants and animals.

The only difference in the description given by the older children is the mention of a landscape (usually woodlands, mountains, or hills).

To the question "If you could send a picture representing Nature, what would you depict?", these were the answers:

Enrico (7): A wood with trees, plants.

Marco (7): All the things that are alive and all the animals.

Eleonora (9): A field with some daisies.

Michele (10): The animals and the plants.

Mariasole (8): Fields of flowers.

Luca (6): Forests, lakes, rivers, fields, mountains, streams.

These answers confirm an idea of Nature considered as a world "other" than the human world.

What Criteria Decide Inclusion in the Natural World?

Group discussion stimulates children to express their opinions and allows them to analyze and justify their ideas after taking those of their schoolmates into consideration. Therefore, the discussion plays an important role in an explanation about what children think of Nature.

One way to facilitate children's discussion is to invite them to make a graphic representation of the ideas as they are taking shape through the exchange of different perspectives. To draw an idea helps one to reflect on it. Therefore, I proposed that the children draw a graphic representation of the meaning they intended to express by the word *Nature* on the blackboard. They were asked to build together the "Nature set" that would gather together all the elements that define the natural world. This was proposed to children in all grades with the aim of (1) checking to see whether a different idea of Nature from that expressed in the first drawings emerged from the discussion, and (2) discovering what criteria were used to decide the inclusion/exclusion of elements.

In both of the younger groups, there were some who, having listed plants and animals of various kinds, proposed to include houses as well. It is evident that for these children the factor that decides inclusion in the natural world is "to be plants or animals."

In the 2nd grades, the suggestion by some to include houses is questioned with another argument: The house is not part of Nature because it's an artificial thing. Many used the distinction between natural and artificial things and decided that only natural things should be included in Nature, that is, those elements "not made by people." This is the criterion that decides the "belonging to Nature" category most used by children in the 2nd grades. If, on the basis of the same criterion, all children agreed to exclude from Nature cars, tractors, and generally all artifacts, some children instead would have liked to include a church even if they recognized that this is also a human artifact:

Pietro (8): Because God is there, and He makes all these things, and so He goes there as well.

In two 2nd grades, several children were much taken up with religious questions. These children showed a complex idea of Nature, a nature inhabited by the divine, without which it is not allowed to exist, an idea which disappears from the thinking of the older subjects.

In all the grades, there were children who proposed the inclusion of heavenly bodies, above all the sun and the stars, an idea which was generally shared. Only in two of the 4th grades was the topic questioned by two children with varying arguments.

One girl in the 1st grade defined Nature as made of "plants and stars." I asked the children if they agreed with this idea, and, in chorus, they replied in the affirmative. One boy added that even the sun and the moon are part of Nature. Since the criterion that decides the "belonging to Nature" used by these children was "to be plants or animals," when I asked why they included the heavenly bodies in Nature, they were obliged to turn to other arguments:

Sara (6): Because the sun has always been in Nature.

Davide (7): The sun gives light.

For children in the 2nd grade, and for most in the 4th, the criterion that establishes whether an element "belongs in Nature" is its "being natural," that is, not made by humans. On the basis of this logic, even what is not on Earth belongs in Nature.

Marco (9): The stars, space, the sun are natural, too.

Michele (10): All of the universe is part of Nature including the planets and the stars because humans don't make them.

These children have a cosmic idea of Nature, not limited to the biosphere. Only two children in a 4th grade class questioned this idea, using a criterion of spatial location:

Marco (10): The sun, the moon and the stars are in space, and it doesn't seem to me that space is Nature. It is the world.

Alessandro made a clear distinction between Nature, the Earth, and the Universe. It is not sufficient that an element fulfills the criterion of "*being natural*" for it to "belong to Nature"; it must also be "on Earth":

Alessandro (10): Nature is on the Earth, and the Earth is part of the Universe, and in the Universe there are stars, which are different from Nature.

To sustain the idea that the sun "belongs in Nature," only one child used a scientific type of argument that revealed knowledge of the cycle of life:

Marco (10): The sun gives light to Nature. There are plants because of the sun. There are herbivores because there are plants they can eat. There are carnivores because there are herbivores. Without the sun, there wouldn't be any forms of life, so the sun gets put in Nature.

The vital relationship between the sun and earthly life makes the sun part of Nature.

Though ecological topics have been dealt with in each 4th grade, only one subject used the information acquired. This leads us to think that scientific knowledge is not seen by these children as relevant in deciding such questions.

The Central Question: Are Human Beings Part of Nature?

In contrast with the preliminary drawings, in which there were no humans, at the beginning of the discussion in both of the 1st grades, one boy suggested that human beings "belong to Nature":

Eros (7): Nature is made of people.

Eduardo (7): In Nature there's grass as well, and us, and fruit.

Only a few among the 6- and 7-year-olds shared this idea, and only four tried to justify their assertions. *Eros (7)*, for example, explained that "people are part of Nature because they live in Nature." But, generally, the pupils in these grades excluded human beings from Nature on the basis of the same criterion that excluded human artifacts:

Elena (7): People aren't part of Nature because they are neither trees nor flowers.

Alice (6): Because they're not plants, birds, flowers.

Ilaria (7): When it rains, people stay at home, and they're not in Nature like the animals are.

Also among the 7- and 8-year-olds, only one spontaneously proposed to include human beings in Nature. To justify his assertion, he used a scientific criterion:

Nicolo (7): I draw a man and a monkey in the "Nature set" because they are always part of Nature. The monkey is an animal, and Man was a monkey a long time ago.

In the 4th grades, the proposal of including human beings in the "Nature set" was made spontaneously by one girl who did not want to draw a subject representing the whole of humankind, but rather an "alpino," that is, a category of people who live in close contact with the natural world. *Alessia (10)* explained that the "alpino" is a member of Nature "because he lives in Nature." It was the only case in which a criterion used to decide inclusion was "to live in direct contact with the natural environment."

Initially, only a few children spontaneously considered human beings as part of Nature. Then the discussion about the question "Are humans part of Nature or not?" promoted a change of opinion in some pupils, and different ideas emerged. If few among the 7- and 8-year-olds supported the idea of human beings belonging in Nature, the older children mostly shared this idea. The most common criterion used to support a human place in Nature was a scientific one: "Since he is an animal, Man belongs in Nature."[2]

Andrea F. (10): I'd put him in . . . because Man is descended from monkeys, and so he is part of Nature like the other animals.

Elio (10): Man is natural because . . . he is like the animals.

Davide (9): . . . because he's made of earth.

Not all agreed to consider humans as animals. Some distinguished human origins from "Man's" present condition, which places him in a different ontological position from the animals.

Davide (7): Once he was an animal.

Sara (7): Man is on one side and the animals on another, because men were monkeys, but then they changed.

Miriam (7): Human beings aren't animals, nor are they things.

The children were asked, "What are the differences between animals and human beings?"

Miriam (7): Man speaks.

Lucrezia (7): Men live in houses and own lots of things, while the animals don't have those things.

Without denying the others' statements, *Marco (8)* said: Men are still living creatures. They are still animals.

This statement found considerable agreement in the class, revealing the presence among 7- and 8-year-olds of a complex representation of the human condition that tries to reconcile the belonging of the human world to Nature with the particular condition of humans, who show the tendency to transcend a given reality for the building of another world, a world of human artifacts.

For others, the criterion used to decide Man's belonging in Nature is the fact of "being natural, that is, not built."

Adelio (8): Man is part of Nature because he can't be built.

But this argument was questioned through religious arguments:

Giulia (8): Man isn't natural; he's artificial...because he was made by God.

This was not the only situation in which the creaturely condition of the human being was assumed as a criterion for deciding about an ontological situation differing from that experienced by other living beings. In the 2nd grades, some children used religious concepts to support their thesis, whereas others contrasted them with the help of scientific arguments (e.g. Man comes from monkeys). Thinking from different presuppositions generated lively discussions from which a rather complex idea of human beings emerged:

Marco (8): Man isn't only natural or only artificial; he's a bit natural and a bit artificial.

Marco also explained that he is artificial since he is born thanks to his parents' wishes: he is not born by chance. This statement shows the intuition of the human being as a particular subject who can escape from environmental determinism in order to act purposefully.

In Bruno's opinion, "to be made by a father and a mother" attests to the fact that the human being is artificial. When asked to explain this idea, some children constructed this explanation: "Human beings make artificial things; we are made by a man and a woman. Therefore, we are artificial." This argument, shared by others, makes it difficult to find the right place of humans, since these children think Man belongs to the "Nature set" anyhow. It is evident that children perceive the complexity of the human condition:

Elena (9): Man isn't a plant and can't be taken for an animal only, partly because he's something more than an animal... he has feelings, intelligence.

Generally, the thesis that human beings are in a special ontological condition is accepted at the moment it appears during the discussion. Some children, however, emphasize that even if humans are able to build a world of artifacts, this doesn't mean they are less natural subjects than other living beings:

Stefano (9): He has something extra, but he's still part of Nature.

And some children (who don't intend to overemphasize the different human condition) point out that "animals, too, have language" and "have intelligence." Others, determined supporters of the idea of a special condition of humans, state precisely, "The animals have instinct. They don't have feelings." Many children support the thesis of a special ontological condition of the human being:

Andrea G. (10): It's true that Man is part of Nature, since he's a living being, but you can't classify him as an animal, because he has a more developed intelligence.

Andrea F. (10): Man makes all the artificial things.

What is interesting is that for many children the criterion that governs Man's "belonging in Nature" is not ontological, but moral.

Roby (8): With all those strange ideas that come into men's heads to ruin Nature, men can't be part of Nature.

The idea that considers "to be part of Nature" as an ontological condition was questioned by children who consider "to be part of Nature" a status acquired through ecologically oriented behavior.

Maurizio (10): Man has the right to belong in Nature only when he controls his desire for wealth and takes care of the environment.

This thesis was argued in three of the four 2nd grades and in all of the 4th grades as a sign of the children's concern about the problem of pollution. Only a few expressed no doubts about Man's belonging in Nature in the remote past of human history. But as soon as Man developed technology, he could no longer consider himself really part of Nature, because, in using that technology, he destroys the environment:

Raffaele (8): Man ought to be part of Nature, because he has animal origins...but today he shows incorrect behavior.

Michele (8): It's true that Man is a natural being, but now, because of his actions, he may not be thought of as part of Nature.

The moral argument has been developed in an articulated way by 9- and 10-year-olds. Many believe a distinction ought to be introduced between people who behave properly, who can be considered part of Nature, and those who pollute and ruin the environment, who are to be excluded.

Andrea (10): Man can destroy Nature when he needs to, but he gets put out of Nature when he does it for fun.

Therefore, "to be a member of Nature" is not an ontological condition; it is a right connected with one's behavior. To acquire this right, one must:

Mariasole (8): Go on foot or use bicycles, not cars.

Valentina (8): Use hoes, not tractors.

Raffaele (9): Not pollute.

Michele (10): Not knock down trees...only when necessary.

According to other children, to be "part of Nature" does not depend on the kind of behavior adopted by human beings.

Sandrea (10): It's not important that Man pollutes or doesn't pollute Nature because he's always part of Nature....When one day men invent a being who is no longer a natural being, that is made like a computer, he'll no longer be part of Nature. But now Man is still part of Nature.

In general, there were many children who emphasized the improper behavior of modern Man toward Nature, revealing a widespread pessimism about the chances of establishing a different lifestyle, above all among 9- and 10-year-olds.

Lorenzo (9): Nowadays men think only of themselves and ruin Nature.

Maurizio (10): Man must save Nature instead of thinking only of money.

Not all were willing to accept a negative idea of human beings as destroyers, and they gave examples of positive treatment of the environment:

Stefano (9): Man doesn't only destroy; he sows as well.

Valentina (10): Man can stay in Nature because he can also grow plants and does other good acts.

Angela (10): Man doesn't destroy all the time. Sometimes he leaves it alone.

Daniele (9): Man protects it; he makes parks. He isn't always bad; he also does good.

In Western culture, a utilitarian mentality prevails that (developed as an anthropocentric worldview) leads us to think of Nature as an instrument, a resource to be used without limits. Implicit in this concept is the presupposition that humans must enjoy a privileged place with respect to Nature, conceived of as an entity endowed with little value. The idea of a natural world without intrinsic value and the idea of the human being as a privileged subject legitimate a boundless use and manipulation of the natural environment.

In contrast with these anti-ecological assumptions, it is essential to develop an "eco-centric" worldview that presupposes an idea of Nature as an entity having intrinsic value, together with an idea of human beings as nonprivileged citizens of the natural world. If it is difficult to believe that the premises for such a radical change in worldview exist in our culture. It may be instructive to read the few and simple words with which a girl expressed the foundations of an ecological way of conceiving the human-Nature relationship:

Elisa (10): In Nature, there are flowers, trees, woods, streams, a lot of clear, water...and Man, too, can enjoy the water. But he must remember that the water is not only for him; it is for the animals, too.

If humans have a "right," it is not that of "using" the natural world as an instrumental good but rather that of being "part of Nature" because it is a good thing in itself, bearing in mind that this right is to be shared with other living beings.

This was the only case in which Nature was perceived as a good not to be used but to be enjoyed together with animals. This idea is fundamental for the birth of a new ecological awareness because it conceives of "Nature as a value" and human beings as part of a biotic community.

How to Represent an Idea: The Children's Images

Much of the discussion was taken up by the research for a graphic representation of the "Nature set" that could render images of the ideas expressed and largely shared. Given the complexity of the arguments, it was not always easy to find a way to represent Man's position in relation to the "Nature set." In order to represent the idea of the complex ontological condition of the human being, belonging either to the natural world or to the human world, a 4th grade child proposed a solution, largely shared, which consisted in building an "intersection set" between the "Nature set" and the "set of artificial things."

Maria (9): I would put [Man] in the intersection because...we are a bit natural and a bit artificial.

Elena (9): I'd make a circle with the things of Nature. Then I'd make a circle with the human artifacts which intersects the Nature set. The area of intersection is the right place for Man, because he is natural but not only natural.

A different solution was proposed by the children who had in a particular way emphasized the idea that Man is a natural being even if he has something more than animals. To explain this, children proposed a subset in the "Nature set" with a notice saying "human world."

Stefano (9): Man has something more, but he is always part of Nature.

In the classes where the moral argument had been amply discussed, the leading position was the one that claims that Man is part of Nature only when he behaves correctly and is excluded when he is destructive. The result of the drawing exercise was to place *inside* the "Nature set" a man walking in the woods and *outside* the set one in a car near a factory.

Raffaele (9): We'll put the bad men in cars and the good ones—let's put them in the midst of Nature.

Michele (10): Let's paint those who pollute brown like thieves.

At the moment of drawing the human figure (usually given a male identity), I asked if one should draw a male or a female:

Martina (9): A female—she's the one who makes the children.

Stefano (10): I'd decide on a man because he's ....

Sandro (9): He's more intelligent.

Marco (10): There are females more intelligent than males and males more intelligent...

Samuele (10): The woman, because the man has the ideas for polluting the air and the water, while the woman takes care of Nature.

Roberto (8): Women pollute as well because they wash with soap and send the soap to the sea.

Ilaria (8): Men pollute more because it's the men who invent machines, tractors, and soap; the women use them.

Alessia (9): In the "Nature set" I would like to put a woman because she looks after animals.

Michele (10): I'd put a woman because it is usually the men who destroy Nature.

Andrea (10): A woman inside because she respects Nature.

These children share the common gender stereotypes of our culture, i.e., the intelligent male, the producer of technology, who invents the instruments for controlling Nature and the idea of Woman as closer to Nature than Man. Significantly, at this age, the females and some males hold a positive idea of women, which shows a move toward the recognition of women's contribution to the building of a human world in harmony with the natural world.

The Thought Thinks Itself

In order to verify the evolution of each subject's position with respect to the arguments that emerged during the discussion, the 9- and 10-year-olds were asked to draw their own idea of the "Nature set" on a sheet of paper and to write an account of their representation.

If human beings were rarely present in the preliminary drawings, almost all subjects later included them, showing that the discussion had brought about a change of perspective. The children were asked to ponder the change in order to understand why at the beginning few had thought to include humans in the natural world:

Maria (9): Usually I don't think of it even if Man is natural.

Sara (10): When I draw Nature, I don't think too much of Man. I think more of Nature...trees and flowers.

Mattia (9): I haven't put Man in because usually he lives in towns, which are all artificial.

Maria (10): Because usually men don't live in the woods.

Beatrice (10): I didn't put Man in because…the word Nature makes me think more of trees and animals.

The children's explanations lead us to think that in our common sense a disjunctive idea of the relationship between Man and Nature exists that tacitly conditions our thinking. The fact that only a process of reflection induces us to think differently about Man's place in Nature confirms the hypothesis that there is a need to promote within ecological education degrees of reflection aimed at probing the ideas and images of Nature and of human beings that occupy our minds.

To develop an ecologically oriented way of thinking, it is not enough to learn the foundations of ecological science. It is also necessary to re-think the metaphysical framework that underlies cognitive activity.[3] It is not enough to develop new ideas and images to acquire knowledge. It is also necessary to re-think our outlooks in order to know the place in which we think (Arendt 1978).

Conclusions

Through drawings and discussions, some ideas have emerged about what children think of Nature and of the quality of the relationship between human beings and Nature. Data obtained in school contexts may not be indicative of the "children's thinking," but of their "school thinking." Perhaps if we lingered to listen to them in other settings (not evaluative and structured, as in school), we might discover other contours of thought. Indeed, the children live out their relationship with the world with a surprising profundity of thinking.

This research revealed the children as deep thinkers who know how to engage their cognitive activities in relevant questions. A weak representation of the child's "life of the mind" ignores the power of her/his thinking which is generally supported by intense emotional participation. It makes us inattentive and unable to listen to the deep doubts and radical questions in which children invest much of their mental energies. To understand how children think, it is necessary not only to find the time to listen to them with an open mind but also to show authentic interest. This is a condition for structuring a conversational context in which the child feels the pleasure (not the duty) to reveal the symbolic world in which she or he thinks.

Notes

1. The children are accustomed to doing this kind of exercise because from the 1st grade they learn the theory of sets and put it into practice.

2. In this passage, "Man" has been retained as the human universal to retain the flavor of the Italian children's thinking.

3. Fundamental questions such as "What is Nature?" or "What is the relation of humans to the rest of Nature?" are identified philosophically as metaphysical questions because they allow for more than one answer (von Foerster 1990).

References

Arendt, H. 1978. *The Life of the Mind.* New York: Harcourt Brace Jovanovich.

Bateson, G. 1979. *Mind and Nature.* New York: Dutton.

Callicott, J. B. 1989. *In Defense of the Land Ethic.* Albany, NY: State University of New York Press.

von Foerster, H. 1990. *Non sapere di non sapere.* In M. Ceruti & L. Preta, eds., *Che cos'e ta conoscenza?* Bari: Laterza.

Leopold A. 1970.*A Sand County Almanac* . New York: Ballantine.

Merchant C. 1980.*The Death of Nature:Women, Ecology, and the Scientific Revolution.* New York: Harper & Row.

Passmore J. 1974.*Man's Responsibility for Nature.* New York: Charles Scribner's Sons.

Chapter 24

Ecological Caring: A Psychological Perspective on the Person/Environment Relationship

Abigail S. McNamee
Lehman College
The City University of New York

Human ecology studies the relationship between people and their environments. The human ecological relationship involves an integration of both cognition and affect, or the thinking and feeling states, in a person. Some of us—maybe many of us—entertain an erroneous assumption that a caring relationship between a person and the environment is solely a cognitive issue and that it could be taught whenever we think it appropriate to insert it into a curriculum.

This assumption is erroneous because environmental caring is not solely a cognitive issue. Nor do I think it can be taught, or, more specifically, that the teaching of environmental caring will "take" unless certain kinds of life experience precede the teaching. The question arises: How did those who care about the environment develop to their current level of caring (however advanced it might be)? How do children become caring? Can we facilitate the development of caring children—the development of care-*full* children, of children who are and continue to develop as people "full of care," rather than as care-*less* people, people "without care" for their environment, be it people, animals, plants, or natural or humanmade objects?

To answer this question, there are some ideas we must consider about relationships. In addition to the conviction that relationships involve both cognition and affect, I would like to mention two more: that relationships develop over time and that relationships, as they develop, are affected by known and unknown changing elements in the person, as well as in the environment. I would like to focus on these aspects of relationships, applying each to the ecological relationship between a person and the environment.

An Ethic of Care Involves Both Cognition and Affect

Our goal in facilitating the development of a caring ecological relationship is double-edged. The first edge is *cognition.* We want a person to understand the complexity of the environment, to understand the necessity and the process of attending to the environment in a caring manner, to understand the mutual dependency between person and environment. Understanding may motivate a person to do what is right, but it may never become a motivator for caring.

The second edge of the double-edged relationship between a person and environment is *affect.* We want a person to feel for his/her environment, to care, to be care-*full*, care-*giving*, rather than care-*less*. This second edge is not likely to develop through an approach which emphasizes the first edge, understanding. Environmental caring is not likely to develop through an objective, factual approach or even a problem-solving approach. Environmental cognition and environmental concern, understanding and caring, may be co-constructed and co-dependent, but that concern cannot be assumed to follow cognition automatically: caring cannot be assumed to follow understanding.

An Ethic of Care Develops over Time.

As a system develops, a history is assumed—a beginning or precursor. When does caring take root and begin to grow in a self-system? When considering environmental caring, it might be easy to assume it is developed primarily in school, perhaps elementary school, but surely not as late as high school. It might also be assumed that environmental caring begins with a good curriculum, in essence an environmental ethics curriculum. I prefer to think caring takes root in infancy, perhaps even in pregnancy, as a mother (and hopefully a father) anticipates caring for the expected baby. It takes root in the anticipatory caring of a caregiver and in the actual caregiving of another person from birth. If it begins there, it took root in the anticipation of the grandparents for the parents, the great-grandparents for the grandparents, etc. It began longer ago than we can return to. It takes root in physical caregiving, accompanied by the psychological caregiving of a nurturing other, and develops into self-nurturing and eventually into the ability to nurture another and to "nurture" the environment. Attachment theory can help us here.

Attachment Theory

Attachment theory is often the focus of debate. There are, however, some generally accepted basics, namely, that every human baby is wired for attachment (Klein 1995) and requires at least one nurturing caregiver to sustain its early physical and psychological development. With human babies, we know that physical care (feeding and cleaning) is not enough; psychological care must also be present and "good enough" for a baby to thrive. This means that the caregiver must be physically and frequently present and must be emotionally connected: gentle in touch and general manner, communicative with eye contact and words, in tune with and responsive to a baby's expressions of need.

While this kind of relationship may be an ideal never again experienced (except at the beginning of a love affair), with human babies, we think we have learned that it is a general atmosphere of being "good enough" that is needed (Stern 1985). An uninterrupted flow of perfect harmony throughout childhood is neither possible nor desired. There can be occasional bad moments, even bad days, between caregiver and child. We also think we have learned that multiple caregivers, substitute caregivers, male as well as female caregivers, work well—with some cautions. They must also be "good enough," they must spend enough time with the child to become attached, and there must not be too many.

With this kind of generalized experience, a baby learns what it means to be nurtured, to be cared for; what it is to feel good physically, cherished emotionally. This experience is internalized by the infant and becomes the core of personality. It becomes the major organizing principle in human experience. Being cared for is the beginning of what it is to care.

Without a reliable or receptive caregiver, a developing child is unlikely to grow into a person who is able to love, able to be care-*full*, full of care. These children are likely to develop into persons who feel unlovable and are unable to love, who are care-*less*, without care, even destructive of self and of their environments. Some children have recently been referred to as "children without conscience" (Magid & McKelvey 1987). They treat the world as they feel themselves to have been treated and rail against an environment that they perceive as withholding in an attempt to exact what they want without being sure of what they really need.

Separation/Individuation

Having experienced "good enough" attachment, a developing child moves on, well fortified, to the second crucial (but often forgotten) aspect of attachment: separation/individuation, or the moving away from dependency on the caregiver. The ideal of early caregiving, which has been essential, would become stifling should it continue. Carol Gilligan writes that attachment and separation anchor the cycle of human life both biologically and psychologically (1982, 151). Separation/individuation requires differentiating the rest of the environment from the caregiver and exploring this environment in larger and larger pieces, as well as with more and more specificity.

Integration of Attachment and Separation/Individuation

The child who has been supported by one or more caregivers in experiencing both attachment and separation/individuation develops a sense of personal wholeness and integrity, a sense of what it is to be *me* and *no one else.* This child can reconcile the need for both oneness with a caregiver and separation from that caregiver. Each reconciliation requires some sense of loss of the preceding modes of organizing the world and the mourning of those losses. This affect is the door both to a full participation in the present and to higher levels of personal and social functioning. Attachment and separation/individuation depend also on environmental circumstances that neither player can control completely (problems small and large in the family, in the community, in the country, in the world).

Both attachment and separation/individuation are co-constructed between a child and whichever caregiver is present and are dependent on the "fit" between child and caregiver. The behavior of real-life caregivers and children falls somewhere on a continuum, not necessarily at an extreme of "good fit" or "poor fit." Many children, perhaps not most, experience "good fit" with their caregivers: there is a synchrony between them, a flow; the personality of each works for the other. Good enough caregiving presupposes a "good fit" that comes naturally and easily or one is achieved through hard work on the part of the caregiver. "Good-fit" children, who are more likely to receive good care, are more likely to grow up into care-*full* people, full of care for themselves, for others, and their environment. They also experience a "good fit" with their environments.

At the other end of the continuum, growing numbers of children experience a "poor fit" that may never be improved: a dysynchrony, a perpetual disruption of flow. A "poor fit" evidences neglect; "poor-fit" children are more likely to grow into people whose loving emotions are split off, who act out rage through destruction. These children may develop as people who are care-*less*, without care for the environment.

In between on the continuum are those children who receive barely adequate caregiving within a barely adequate "fit" with caregivers who are barely present physically and/or who are barely present emotionally, caregivers who are rather casual in their caregiving, only partially connected. Such children are more likely to split off their emotions and to withdraw from the felt need for either care-receiving or care-giving and, in so doing, may avoid intimate relationships with the people, animals, plants, and natural and humanmade things in their environment.

Readiness for Ecological Caring

Children who receive *good enough* care in a "good-fit" relationship with their caregiver/s are primed and ready for the environmental teaching we offer at home and in our schools. These children will be most able to integrate their emotional and intellectual selves and develop into people who truly care about the environment.

Children who receive *barely adequate* care within a barely adequate "fit" relationship with their caregiver/s may be somewhat ready for environmental teaching. These children are likely to learn "the right thing" or the "just thing" and do it, but compulsively and without emotion. They may be interested in the rules and follow them. They may understand environmental needs and act on them, but they won't be touched emotionally. It is also possible that these children will be indifferent to their environment and to environmental teaching and will withdraw from environmental caring but without being destructive. They will be ecologically neutral. At best, we may reach them intellectually with a spectacular curriculum. Rarely will we reach them emotionally.

Children who receive *poor* care within a "poor-fit" relationship are the most likely to remain untouched by whatever we offer in the way of environmental teaching at home and in school. Their rage over not having gotten what they needed when they needed it will make it impossible for them to connect with teachers, as it was impossible for them to connect with parents. They may find it hard to connect with nature or their environments. They will be unlikely to conceptualize or feel themselves as care-*givers.* Their rage will prevent care-*full* behavior. They may act "as if" our environmental teaching is interesting or impactful, but the goal is to protect themselves from further harm, and to accomplish this, they may turn to deceptive charm, to blatant deception, or to destruction–including environmental destruction.

In the extreme, these are the children who kill or torture animals repeatedly. They have been known to kill other children "to see how long it takes them to die" (Magid & McKelvey 1987), the result of rage and curiosity without care. These are the children who killed a 2-year-old in England a couple of years ago. These are the adolescents who put a live fish in a microwave in Florida "to see what would happen," the same adolescents who put a puppy in a bag and blew in marijuana "to see what a puppy was like when it got high." These are the adults in New Jersey who applied to adopt a Dalmatian from an aging farm couple and tortured it, cutting off its ears and its tail and cutting open its stomach while it was still alive. These are the adult policemen in New York who answered a call to pick up a found Shi-tzu dog and took it to a field for gun practice. These are the children for whom EE will fall on deaf ears and hardened hearts despite our best efforts. Such children have been damaged, perhaps beyond repair, by what came before in their lives, before we ever meet them as teachers.

Throughout the world, well-nurtured children are becoming concerned about, and often actively involved in, caring for the people, animals, plants, and natural and humanmade objects with which they interact in their environment (see Mortari, this volume.) Schools are taking an increasing role in teaching ecological caring, in encouraging all children, however nurtured, to investigate, plan for, monitor, and manage the environments of their own communities (Hart 1994). The teaching strategies utilized in EE assume that children can develop an environmental ethic, or Earth stewardship, but often ignore the developmental route that children move through toward an ever more mature ethic of caring.

Feminist Contributions to an Ethic of Care

The work of Carol Gilligan (1982, 1988) and others (Larrabee 1993; Lyons 1993; Thompson 1992, 1994; Tronto 1993) is particularly relevant to environmental caring. These theorists recognize and call attention to the fact that many have defined the female personality by using male personality as the norm.

It became clear to Gilligan (1982) that women's development was located in their experience of relationships, that feminine personality defines itself in relation to, and in connection with, other people more than does masculine personality. Women define themselves in a context of human relationships and judge themselves "in terms of their ability to care" (17), which determines their construction of moral problems. For Gilligan, the standard of moral judgment that informs women's assessment of self is a standard of relationship, an ethic of nurturance, responsibility, and care. Women measure their strength in the activity of attachment, i.e., "giving to," "helping out," "being kind," "not hurting"(159).

Gilligan found that in all of women's descriptions, identity is defined in a context of relationships and judged by a standard of responsibility and care. She says that "morality is seen by these women as arising from the experience of connection and conceived as a problem of inclusion rather than one of balancing claims" (160), the underlying assumption being that morality stems from attachment.

Joan Tronto (1993) adds that an ethic of care cannot be applied equally to all situations. People do not care for everyone and everything equally; "it is easy to imagine that there will be some people or concerns about which we do not care," but does lack of care "free us from moral responsibility" (249)? An ethic of care must also be "situated in the context of existing political and social theory" and may constitute a view of self, relationships, and social order that is incompatible with an emphasis on individual rights (251). She asserts that an ethic of care is "a set of sensibilities which every morally mature person should develop, alongside the sensibilities of a justice morality" (252). This implies that caring can be taught in families and schools and that males and females have equal potential to develop both.

Nona Lyons (1983) and Patricia J. Thompson (1992, 1994) attempt to remove an ethic of care from gender division. Lyons defines two modes of describing the self in relation to others: separate/objective and connected. In making moral decisions, separate/objective individuals (of either gender) "tend to use a morality of justice," while connected individuals (of either gender) tend to use "a morality of care." Each construction has strengths and weaknesses. Equality is an ideal and the strength of a morality of justice; consideration of an individual's particular needs is an ideal and the strength of a morality of care. Impartial concern for the rights of others may not be enough to provide for care, and caring for others may be overly emotional or even unfair (135).

Thompson (1992, 1994) describes not only a different voice but also a different location or social position as generating two modes of orienting the self in relation to others and to the environment in two systems of human action. She identifies the Hestian/Hermean perspectives, which are not divided on the basis of gender. The Hestian perspective emanates from the *oikos* (the private domestic sphere) and operates according to an ideology of connection, on an ethic of care, and is reinforced by intrinsic rewards. The Hermean perspective emanates from the *polis* (the public civic sphere) and operates on an ideology of control, on an ethic of justice, and is reinforced by extrinsic rewards. The environment, too, can be viewed either from a Hestian perspective of connection and care or from a Hermean perspective of control and domination.

Conclusions

An ethic of care is the foundation of a caring ecological relationship and, as such, is dynamic and can be affected by the internal environment of the person and by the environment in which he/she lives. I would like to conclude with this aspect of an ecological relationship, using it to move toward at least a temporary integration of the influence of attachment on the development of a caring relationship with the environment. The notion that a caring ecological relationship is dynamic implies that it is an open system influenced by internal and external environments. Our internal and external environments consist of the interaction of biological levels (molecular, physiological, individual, group/population) with the multiple social and psychological influences over time of everything that surrounds us. Our internal and external environments, in all their complexity, affect each other and, of course, us. Interwoven through these are those two influences we have called "nature" and "nurture," which cannot wisely be separated as explanations for human behavior. We are dealing with people who are dynamic organisms, who are in some kind of dynamic relationship with their dynamic environments. This dynamism makes for remarkable complexity, which we try to organize and order.

Let us return to the specific questions raised in this chapter: How did those who care about the environment develop their current level of ecological caring (however advanced it may be)? How do children become caring? Can we facilitate the development of children who are, and continue to develop, as people "full of care" rather than as people "without care" for the environment?

To answer these questions, we must recognize that we are groping with complexity in both the questions and the answers. Attachment theories and morality theories each offer insight but must move toward further integration within the theories themselves and between the theories. It seems apparent that the experience of early and continued attachment relationships affects the development of an ethic of ecological care and must precede any effort at teaching EE. Recent research on infancy provides compelling indication that the foundations of morality are present early in child development (Gilligan & Wiggens 1988; Kagan 1984; Stern 1985) both in an infant's responsiveness to the feelings and care of others and in the young child's appreciation of standards (Gilligan & Wiggens 1988, 114). It also seems apparent that an ethic of ecological care involves both the cognitive and emotional aspects of environmental stewardship and that, accordingly, both are worthy characteristics in all people regardless of gender. It remains for developmental theorists, educators, and parents to integrate their thinking and action so that we can continue to move forward in conceptualizing and facilitating the development of people who are care-*full* rather than care-*less* of the environments we share.

References

Chawla, L. & R. Hart. 1995. "The Roots of Environmental Concern." *NAMTA Journal. 29*:1 (Winter), 148-59.

Eyer, D. E. 1992. *Mother-infant Bonding.* New Haven: Yale University Press.

Gilligan, C. 1982. *In a Different Voice.* Cambridge, MA: Harvard University Press.

Gilligan, C., J.V. Ward & J.M. Taylor, eds. 1988. *Mapping the Moral Domain.* Cambridge, MA: Harvard University Press.

Gilligan, C. & G. Wiggens. 1988. *The Origins of Morality in Early Childhood Relationships.* In Gilligan, Ward & Taylor (above), 111-38.

Hart, R. 1993. "Affection for Nature and the Promotion of Earth Stewardship in Childhood." Keynote lecture for The American Horticultural Society's National Children's Gardening Symposium (August).

Hart, R. 1994. "Children's Role in Primary Environmental Care." *Child-hood 2*, 92-102.

Hart, R. & L. Chawla. 1981. "The Development of Children's Concern for the Environment." *Abhandlungen. ZfU 2/81*, 271-94.

Kagan, J. 1984. *The Nature of the Child.* New York: Basic Books.

Kaplan, L. 1984. *Oneness and Separateness: From Infant to Individual.* New York: Simon & Shuster.

Kavaler-Adler, S. 1992. "Mourning and Erotic Transference." *International Journal of Psychoanalysis 73*: 3, 527-39.

Klein, M. 1940/1975. "Mourning and Its Relation to Manic-depressive States." In *Love, Guilt and Reparation and Other Works 1921-1945.* London: Hogarth.

Klein, P. F. 1995. "The Needs of Children." *Mothering* (Spring), 39-45.

Larrabee, M. J. 1993. *An Ethic of Care.* New York: Routledge.

Lyons, N. P. 1993. "Two Perspectives on Self, Relationships, and Morality." *Harvard Education Review 53*:2, (May)125-45.

Magid, K. & C. A. McKelvey. 1987. *High Risk Children Without a Conscience.* New York: Bantam.

Mahler, M., F. Pine & A. Bergman. 1975. *The Psychological Birth of the Human Infant.* New York: Basic Books.

Stern, D. 1985. *The Interpersonal World of the Infant.* New York: Basic Books.

Thompson, P. J. 1992. *Bringing Feminism Home.* Charlottetown: Prince Edward Island, Canada: Home Economics Publishing Collective.

_____.1994. "Hestian Ethics: The Question of *Byt* in the Daily Lives of Russian and American Women." Paper presented at conference on "Social Changes in East and West: The Public of Women–Women's Research–Women's Policies II," Akademie Frankenwarte: Friedrich Ebert Stiftung, Würzburg, Germany (July 9-16).

Tronto, J. C. 1993. "Beyond Gender Difference to a Theory of Care." In Larrabee (above).

_____. 1992. *Moral Boundaries: A Political Argument for an Ethic of Care.* New York: Routledge.

Wolf, L. C. 1975. "Children's Literature and the Development of Empathy in Young Children." *Journal of Moral Education 5*:1, 45-49.

Chapter 25

Diane Ackerman's Niche in Writing "Natural History"

Gene McQuillan
Kingsborough Community College
The City University of New York

In 1990, John Tallmadge declared, "Today, with writers like Barry Lopez, Annie Dillard, and Robert Finch at work, nature writing is arguably the most exciting realm of American literature" (64). Jay Parini's 1995 article, "Greening the Humanities," opens by stating that "Deconstruction is compost. Environmental studies is the academic field of the 90's" (52). The article further explains that nature writers and practitioners of ecocriticism enjoy a privileged status on certain university campuses, and it portrays them as having an enviable off-campus life as well. Parini begins by describing a trip during which he meets what seems to be a planeload of East Coast professors heading to a conference in the Colorado Rockies sponsored by ASLE (Association for the Study of Literature and the Environment); the facing page of his article displays a photo of the writer/professor John Elder paddling calmly in his homemade cedar canoe. At times I have found myself in hearty agreement with writers such as Tallmadge and Parini, and I cannot help but become envious of the professors they mention. Yet more often I recall how defensive I become while arguing the merits of nature writing with friends, enemies, and colleagues. Some people read the "manifestos" of Tallmadge and others as little more than preaching to the converted; others demand to know why nature writing should be considered more "exciting" than feminism, cultural studies, or virtual reality. More than a few people express doubts about the "American" characteristics of nature writing, and others need to be convinced that nature writing even deserves to be called "literature." As Karen J. Winkler has pointed out in "Inventing a New Field: the Study of the Literature About the Environment,"

Parini received over 100 letters about his article—"most of them angry"
(A9).

In fairness to such responses, it should be mentioned that such doubts
circulate even among active supporters of nature writing. Proponents of
nature writing often confront this basic point of resistance: nature writing
is by definition "about" various relations to and within natural environments, and since that source is damaged, contemporary writers are basically reduced to humble claims or angry elegies for the "diminished thing"
known as the American landscape. Instead of the full-throated songs of Muir
or Emerson, the modern nature writer is limited to polite humming or protest slogans. It is indeed tempting to argue that nature writing would be an
unlikely product of post-industrial American culture. There are abundant
signs that "the fruited plain" has been reduced to a "Garden of Ashes," and
many American writers have confirmed this fear. Even the most fervent supporters of wilderness and nature writing, such as Roderick Nash in *Wilderness and the American Mind* (1980), will admit that: "The blank spaces are
being filled in. Today, not 1890, is the real end of the American frontier"
(380). Thus, it is not surprising to hear claims that writings about Nature
and wilderness in 20th-century America have been a well-intended but fu-
tile effort to recover the immediacy and the complexity of those writers—
Bartram, Lewis and Clark, Cooper, Catlin, Thoreau, Fuller, Melville,
Parkman, Whitman, Twain, Muir, Cather, Leopold—whose texts rely on
direct experiences of the American landscape, whose visions of America
drew upon rich and sustaining beliefs in the regenerative powers of nature.
Until recently, one could find this idea expressed almost automatically in
literary criticism or history that focused upon the legacy of American nature and wilderness; in general, scholars have aligned the "closing" of the
frontier with a final closure of certain literary possibilities. Contemporary
criticism about nature writing has often tried to counter such accusations
and to articulate the ways in which a revised understanding of the genre of
nature writing leads to a rethinking of its place within American culture.

Exhausted Genres

There is no shortage of nightmarish descriptions of the modern American
landscape; however, the arguments for the narrowing prospects for nature
writing involve difficulties more subtle than Three Mile Island, the Exxon
oil disaster, or the Hanford nuclear waste site. Some of the most significant

arguments concerning the decline of nature writing have been proposed in John Hildebidle's *Thoreau: A Naturalist's Liberty* (1983) and Sherman Paul's *For Love of the World* (1992).

One common thread in arguments "against" nature writing is the explicit sense that nature writing after 1890 is a tired genre. Hildebidle states that he is "most interested in natural history as a literary method and indeed as a literary genre." He delineates the methods and purposes of 19th-century natural historians and shows how their texts are guided by three crucial ideas, namely, the principle of inclusiveness—the refusal, that is, to turn away from any source of information or any mode of discourse; the centrality of personal observation; and the clear attachment to place. He then proceeds to read several of Thoreau's texts—"Dispersion of Seeds," "Wild Apples," *Walden, Cape Cod*—to analyze how Thoreau adapts these strategies as a means of unifying the almost unlimited scope of his interest in the natural world. A reader of *A Naturalist's Liberty* is likely to be convinced that Robert Frost had the right idea when he said that nature writing should not just be about "pretty scenery" but about the "Whole Goddamn Machinery." Hildebidle first aligns nature writing with "the principle of inclusiveness," with a generous attention to history and geography, myth and literature, science, and folklore. Yet he seeks not only to praise Thoreau but also to write the epitaph for nature writing after Thoreau. Those who might have been thinking of the "inclusiveness" of Barry Lopez's *Of Wolves and Men*, Aldo Leopold's *A Sand County Almanac*, or Annie Dillard's *Pilgrim at Tinker Creek* should know that Hildebidle's discussion has been framed by the following statement: "It is my argument that Thoreau is in fact the last— and in literary terms, overwhelmingly the most successful and important— of the natural historians" (1983, 5). Hildebidle places Thoreau's texts directly within what he calls a "sea-change" in the methods and purposes of natural history writing. He implies that the far-reaching skills of Gilbert White or Thoreau have been replaced by the narrow disciplines of the specialist: the wildlife biologist, the travel writer, the mountain guide. Hildebidle also insists on the "centrality of personal observation," but he attaches some strict conditions to this basic act of observation. He argues that during the mid-19th century a breach had opened between the logic of applied science and the literary tendency to see nature as an Emersonian symbol of spirit. Once again, Thoreau is seen as the last to unify these tendencies: "Thoreau is arguably the last major American writer to believe that he could be both a scientist (in some actual, nonmetaphorical sense) and a man of letters, at the same time and in the same work" (95).

Hildebidle's claims help to explain a recurring problem faced by contemporary natural historians, who find that they are often not "allowed" to maintain a dual allegiance to science and literature: while it is problematic to define the theoretical boundaries between "scientific" and "humanistic" discourse, such distinctions are often made in practice by anthologists, publishers, awards committees, and college programs. While many readers are fascinated by Stephen Jay Gould's writings on natural history, I do not expect to read these essays in the next *Heath Anthology of American Literature*. John McPhee's *Encounters with the Archdruid* (1972) was nominated for a National Book Award in the category of science—yet I think it far more likely that McPhee would be considered a man of letters than a scientist. It is also less likely that a contemporary writer would have the "clear attachment to place" that Hildebidle lists as his third criterion. There are fewer and fewer Walden Ponds to serve as private sanctuaries, and the public domain of national parks and wilderness preserves often fails to inspire recent writers. In general, Hildebidle's distinctions effectively displace natural history into the margins of literary history. In the 19th century, natural history was a relatively major genre; in the 20th, it has often seemed an isolated curiosity, a poor relative of both literature and science.

Hildebidle has written one of the more important texts about 19th-century natural history, yet problems begin when he discusses texts written after 1880, for then he shifts to a far more limited perspective. Instead of portraying the development of a particular concept of nature or nature writing as it occurs in a wide variety of texts, he refers to a sudden sense of limitation among writers and proceeds to rely upon a relatively narrow range of texts to support such assertions. When Hildebidle refers to Thoreau as "the last natural historian," he assumes our assent rather casually. Sherman Paul raises such a question in his review of *A Naturalist's Liberty* (1983):

> For one thing, it might help to explain what happened to the genre that Thoreau found so congenial and so wonderfully adapted to his own ends. Did it die with the advent of professional science? Isn't there a tradition of nature writing that extends beyond John Burroughs, the last in the Thoreauvian line mentioned by Hildebidle—the tradition in which such diverse practitioners as E. B. White and Annie Dillard, Edward Abbey and Sigurd Olson are enrolled? (461)

Paul tries to answer his own question in *For Love of the World: Essays About Nature Writers* (1992), which should be considered among the most subtle and informed responses to the criticisms that Hildebidle and others have made about contemporary American nature writing. Of course, Paul is best known to many readers as a scholar of Thoreau's work, and he begins *For Love of the World* with a trio of essays about him. His claims about Thoreau become the underpinning for many of his later arguments, and they also lead Paul to a serious conflict about the social purposes of contemporary American nature writing. Paul is not afraid to admit when he is still unsure about difficult questions, and one crucial dilemma he ponders is how well recent writers match up to their predecessors:

> Nature writers—I'm thinking of many of the contemporary writers represented in the Antaeus collection (1987), edited by Daniel Halpern and *Words from the Land; Encounters with Natural History Writing* (1988), edited by Stephen Trimble—are "fine" writers. As Trimble says, those he has selected see themselves as writers first. They are personal essayists, often of distinction, who find some of their material in nature, and many of them sit for their own portrait when they go there. Seldom are they naturalists like Thoreau, Muir, and Leopold, who articulated a profound, lifelong engagement with nature, more often wrote to live than to make a living, and preferred advocacy to pacification. Taken together, the *Antaeus* collection and *Words from the Land* give us a canon and establish nature writing as fashionable, a genre ready-to-hand for writers' workshops. (1992, 68)

Sympathetic Reunions: The Contributions of Diane Ackerman

Since 1983, Diane Ackerman has accumulated a list of publications, teaching assignments, and travel destinations that could make Jay Parini's planeload of ASLE professors seem downright provincial. A five-part PBS series featured her writings and commentary, and these appearances could perhaps be considered as the culmination of praise for this contemporary nature writer. In 1990, Ackerman published *A Natural History of the Senses*, which became a national best-seller. In the same year, her article "Nature Writers: A Species unto Themselves" appeared on page 1 of the *New York Times Book Review*. For the moment, I will skip past Ackerman's public persona to focus instead on some of the ways in which her recent writings confirm and challenge some of the criteria of scholars such as Hildebidle and Paul.

It has previously been mentioned that Hildebidle defines three aspects of natural history writing that he considers crucial, the first being "the principle of inclusiveness, the refusal to turn away from any source of information or any mode of discourse." Ackerman begins *A Natural History of the Senses* by stating that "there is no way in which to understand the world without first detecting it through the radar net of our senses," and a quick browse through the table of contents alerts us to all the things that our radar might hone in on. Chapters include "Notions and Nations of Sweat," "Adventures in the Touch Dome," "The Psycho-pharmacology of Chocolate," "Is Music a Language?", and "How to Watch the Sky."

One reason for Ackerman's success with a general reading public is that she doesn't apologize for indulging her senses. Many nature writers focus attention on the senses by stripping away anything that might distract them from the direct perception of *that* birch leaf, *that* thrush music, *that* slant of light on a bone in the desert. Muir went into the Sierras with little more than a tough coat and a tougher loaf of bread; Thoreau made lists of the basic necessities he had done away with; Edward Abbey floated down the Colorado River with a toylike rubber raft, no life jacket, some cans of beans, and some pipe tobacco; Sherman Paul's shack on Wolf Lake is almost as bare as Thoreau's. Ackerman's trips lead to perfume houses, storerooms full of spices, temples of massage, laboratories of sensory extremes, orgies of eating and bodily delights, circuses of odd customs. One does not find any Wordsworthian huts in her texts; Blake's dictum "The road of excess leads to the palace of wisdom" seems closer to her guiding principle. She revels in letting us in on the lightest touch, the brightest light, the "Limits of Hearing, The Power of Sound." She also keeps readers off-balance with references to the foulest smells (the open sewers of Paris in the late 18th century) and the strangest recipes (including those for stir-fry puppy and for roasting a goose while it is still *alive*). Of course, such indulgences can often lead to either careless writing or vulgar appreciations, but Ackerman prefers to challenge us with gentle reminders of our complacent responses to the natural world. In "How to Watch the Sky," she begins by asking us where the sky is, and as our minds float upwards toward the ceiling, she simply tells us, "Look at your feet. You are standing in sky." A reader learns the names of winds in 19 different languages, reads myths of the creation of the sky, catches glimpses of hawks and hang gliders, charts the stars in different hemispheres, and contemplates the crescent-studded skyline of Istanbul. *A Natural History of the Senses* is nothing if not "inclusive."

The 14 pages of "How to Watch the Sky" form the longest chapter of *A Natural History of the Senses,* which has been consumed by many like a bizarre box of holiday chocolates. The text features its share of technical vocabulary, but Ackerman has a knack for including unusual details or intriguing incidents that clearly convey scientific developments without trivializing them. When it comes to the "principle of inclusion," Ackerman excels at recreating worlds of sensory experience while also reminding a reader that the late-20th-century naturalist does not need to make any apologies to a 19th-century predecessor when it comes to a range of information about the natural world. The table of contents should be understood almost literally, but as Ackerman loads the table with samples, you do not get a chance to *look* at them. Ackerman shares Emerson's opinion that vision is the least fertile of the senses, so the book encourages one to smell, to touch, to taste, to listen, and then finally to remove the blindfold.

Of course, any book for sensualists won't insist on too much restraint, and one obvious pleasure of reading—and teaching—the book is that there are no necessary sequences for doing so. Almost all chapters are less than ten pages, and many are just two or three, so one may read it with a very 20th-century attention span. This sense of pacing becomes even more noticeable, yet much less satisfying, when the book is seen in its TV form. In some ways, one would expect the opposite reaction, since TV obviously depends on such short segments of information. Yet the TV version cannot manage the intricacy of detail and the wry turns of language that enliven so many parts of the book. Actually, many of the descriptions during the TV segments are taken directly from the book, but the immediate presence of visual cues—a picture of a massage, as well as a verbal description—displaces one's attention from word to image.

Hildebidle also insists on "the centrality of personal observation," and Ackerman clearly enjoys telling a reader how it feels to rub a penguin's tongue, to eat dangerous foods, to fly a small airplane, or to reach one's arm into the birth canal of a cow: "Up to the shoulder inside a cow, you feel the hot heavy squeeze of her, but I'll never forget my startled delight the first time I withdrew my hand slowly and felt the cow's muscles contract and release one after another, like a row of people shaking hands with me in a receiving line" (81). However, it soon becomes clear that Ackerman, while familiar with developments in psychology, biology, anthropology, and related fields, is not a scientist, but a very well-read and astute *student* of science. Thoreau devoted his later years and later writings to an extensive study of the dispersion of seeds; Muir used his gifts as a climber as the basis for developing theories of glaciation; contemporary nature writers rarely find themselves in a position to make such discoveries, but instead find themselves in the position of a spokesperson to a general audience. It isn't that Ackerman doesn't try. Much like John McPhee or Barry Lopez, she has a keen talent for finding scientists who welcome writers into their worlds.

Perhaps the strangest effect of reading *A Natural History of the Senses* has to do with the "clear attachment to place" that Hildebidle listed as the third crucial aspect of natural history writing. In simple terms, Ackerman represents a sort of "super-tourist." While reading *A Natural History of the Senses,* one begins to wonder if Ackerman ever goes home. At times, her writings exemplify the perpetual restlessness that Elizabeth Bishop spoke of in the poem "Questions of Travel" (1965). Ackerman's trips can easily be understood as searches for "the strangest of theaters," "the tiniest green hummingbird," "some inexplicable old stonework"—in Bishop's terms, she seems "determined to rush."

In the chapter "How to Watch the Sky," Ackerman adopts a familiar descriptive technique, that of watching a place, be it a pond, a cathedral, or a view of the sky, for one day. But by noon she is driving four hours south along the Big Sur coast for other perspectives. When Elizabeth Bishop referred to "home," she quickly added "wherever that might be." While "home" is among the most elusive of words, one can think of many writers, be they adventurers, nature writers, photographers, or activists, who developed at least some "clear attachment to place." Think of John Muir and "the Range of Light," Aldo Leopold and his "sand counties," Leslie Marmon Silko and the mesas of the Laguna lands, Anne LaBastille and the Adirondacks, Georgia O'Keeffe and Ghost Ranch, Edward Abbey and the canyonlands of Utah, Ansel Adams and Yosemite, or Rick Bass and Montana. There is no such place with which one can link Ackerman.

At least one of my colleagues who read the book was surprised to hear that I thought of this wide-ranging book as "nature writing." In her article on "Nature Writers: A Species Unto Themselves," Ackerman drops numerous hints about her own affiliations:

Although nature writers vary in their motives and their habits, they tend to have much in common. *The Norton Book of Nature Writing* (1990), a hefty new anthology, is really a sort of village that sprawls across time rather than the countryside. Although few of its inhabitants have ever met, they have hunted in the same light-dappled woods, learned from the same masters, angled in the same streams of thought, and drunk from the same well, whose rim is the horizon. The fences that divide them are low-lying and arbitrary—a century, a nation. Spiritual neighbors, as well as contemplative wayfarers, they would probably be the first to admit that every sympathetic reading of another writer is an act of reunion. (42)

Sherman Paul asked whether there is a tradition that continues beyond John Burroughs, and Ackerman clearly asserts that there is. More importantly, she does not seem overly concerned about the desire—or what some perceive to be the desire—to match or exceed the standards of earlier writers. These distinctions, she claims, are "low-lying and arbitrary," and the love of nature is seen to be a force that would erase or at least mitigate the influence of "a century, a nation" on a writer—or on the new readers who come to appreciate nature writing. She speaks of a "reunion," and the "sympathetic readings" she celebrates can be found in the daily love of writing and reading, as well as in more formal appreciation, such as literary anthologies, PBS specials, and ASLE conferences.

At times Ackerman's writings represent the world of ultra-modern tourism, of a super-traveler armed with an expense account, an open-ended plane ticket, a series of "immodest demands" that would make a pirate blush, and a hell of an attention deficit. (Don't expect to see her on TV wearing "wild wool.") At her best, however, her writings recover, at least for a moment, the sense of wonder that makes order out of mere restlessness, that encourages a reader to respect—and play with—the abundance of the natural world. Ackerman expresses great faith in "visions," and I think this is a crucial phrase for understanding why collections such as Thomas Lyons' *This Incomparable Lande* (1989) have only begun to mine the possibilities of writing about travel, adventure, or landscape. Literary history has almost always been a history of "bright suns," but recent scholars have become just as interested in the "dark suns," whose presence is definite yet still unrevealed. This process of revealing and of reveling, or at least the faith that such a process is possible, is at the heart of all efforts to discover.

References

Ackerman, Diane. 1990. *A Natural History of the Senses.* New York: Random House.

_____. "Nature Writers: A Species Unto Themselves." The *New York Times Book Review* (19 May), 1+.

Bishop, Elizabeth. 1965. *Questions of Travel.* New York: Farrar, Straus & Giroux.

Finch, Robert & John Elder, eds. 1990. *The Norton Book of Nature Writing.* New York: Norton.

Halpern, Daniel, ed. 1987. *On Nature: Nature, Landscape, and Natural History.* San Francisco: North Point Press.

Hildebidle, John. 1989. *Thoreau: A Naturalist's Liberty.* Cambridge: Harvard University Press.

Lyons, Thomas. 1989. *This Incomparable Lande: The Book of American Nature Writing.* Boston: Houghton Mifflin.

Nash, Roderick. 1990. *Wilderness and the American Mind.* 3rd Edition. Yale University Press.

Parini, Jay. 1995. "Greening the Humanities." *The New York Times Magazine.* (29 October); 52-3.

Paul, Sherman. 1992. *For Love of the World: Essays on Nature Writers.* Iowa City: University of Iowa Press.

Tallmadge, John. 1990. [Rev. of] *The Norton Book of Nature Writing* and *This Incomparable Lande. Orion Nature Quarterly* 9 (Summer), 63-4.

Trimble, Stephen, ed. 1988. *Words from the Land: Encounters with Natural History Writing.* Salt Lake City: Gibbs M. Smith.

Winkler, Karen J. 1996. "Inventing a New Field: The Study of Literature and the Environment." *The Chronicle of Higher Education* (August), A8.

Chapter 26

Education for Ecological Literacy

N. J. Smith-Sebasto
University of Illinois
Urbana-Champaign

I magine an extraterrestrial (I'll call it ET) observing Earth, having been sent out by its leaders to search for life in the universe. (As in so many science fiction movies, pretend it looks relatively humanoid and speaks your native language). After careful observation and consideration, ET concludes that humans are the dominant life form on Earth and attempts to establish contact. ET picks an ordinary adult from the population (here limited to my homeland, the United States) and indeed establishes a communication channel.

ET asks this average adult (one who has completed 13 years of formal, structured education) to talk about life on Earth. The dialogue might go something like this:

ET: Tell me about life in your world. What, for example, is this thing under which we're sitting?

Human: It's called a tree.

ET: What kind of tree?

Human: I don't know. It's just a tree.

ET: Well, what about that thing over there. It looks like a tree. Is it?

Human: Yes, it is also a tree.

ET: But it looks different from this one. What kind of tree is it?

Human: I don't know. It's just another tree.

As time goes on, ET and the human continue to explore life on Earth. ET's appetite for information is insatiable. Gradually, the seasons begin to change. ET wants to know about this since this does not occur in its world.

ET: Why are the leaves on the trees changing colors?

Human: Because autumn is coming.

ET: It's also getting cooler. Why?

Human: I don't know, but I think it's because we're closer to the sun in the summer and further away from it in the winter.

ET: I get the feeling you do not know a lot about life in your world!

Human: Well, I don't really need to know that stuff. I know there are some nature freaks and environmentalists who know more about it, but it's not real-world stuff. I mean, what good would it do me to know how to identify trees or which plants I can eat or which ones have medicinal properties? Who cares what causes the seasons? There's nothing I can do about it, except maybe move to where it's warm all year long. I know where the food store is, how to drive a car, and how to balance a checkbook. But don't get me wrong. I care about the environment. I have a recycle bin at home! Anyway, I'm too busy with the important things in life to worry about that stuff. We should go in now. It's almost time for some TV.

After their visit, the human and ET wave goodbye. ET returns to its world; the human, to its.

Environmental Literacy Through Nature Study

The point of this story is that I believe if this scenario were really to occur, ET would likely leave the contact painfully unfulfilled and dissatisfied with what it learned about life on Earth. If ET didn't contact an environmentally literate human, it would leave Earth without the knowledge that would help it fulfill its mission. The type of environmental literacy I am suggesting is principally developed through nature study. As Anna Botsford Comstock (1994) suggested, nature study "makes a learner familiar with Nature's ways and cultivates her/his imagination, cultivates a love of the beautiful, but, more than all, gives a sense of companionship with life out-of-doors and an abiding love for Nature" (1-2). This type of education, which is still common among many present-day aboriginal cultures, was common to many of the early settlers of North America, including the inhabitants prior to the Europeans. Such education, which remained an elusive part of Western formal education until the efforts of Comstock, is today once again nearly nonexistent in formal education. It is, to me, the single greatest failing of the discipline called environmental education (EE).

> Environmental education...must be far broader than words on a page or images on a screen or even classroom learning. Thousands of Americans live far from the natural world, surrounded by the concrete and steel of cities or by the clapboard and cul-de-sac of suburbia. Those who do live close to the land too often have had limited exposure to the concept of stewardship of the [E]arth's resources. (Flicker 1996, 6)

Richard Louv (1990) reported that two years of traveling the country, talking with and listening to parents, children, and, upon occasion, experts, left him with stunning, sometimes terrifying perceptions of parents and children describing an environment that no longer makes much sense—a divorce from nature, an environment that no longer nurtures children (52).

As David Orr (1992) has suggested, "[nature study] is concrete and requires direct involvement in nature. It requires firsthand knowledge of [plants and] animals," both terrestrial and aquatic, as well as Earth systems and Earth science. "[Nature study] forces us to deal with nature on nature's terms. It promotes the capacity not only to see but to observe with care, understanding, and, above all else, with pleasure" (136). He goes on to suggest that such "careful observation" is missing and uses as an illustration an "environmental perception of place" test published in *Co-Evolution Quarterly*. The test items include:

1. Trace your drinking water from its source to your tap.
2. What kind of soil are you standing on?
3. What was the total rainfall in your area last year?
4. When was the last time a fire burned your area?
5. How did people live (agriculture/farming) in your area before you?
6. Identify five native edible plants in your region and their season(s) of availability.
7. Where does your garbage go?
8. Name five grasses in your area. Are any of them native?
9. Name five resident and five migratory birds in your area.
10. What species have become extinct in your area?

(Adapted from Charles, Dodge, Millman & Stockley 1981-1982, 1).

Orr suggests that the fact that few people would be able to correctly answer many (if any) questions highlights a widening gap between the growing power of our society over nature and the general ignorance about it among individuals. Equally surprising is the acceptance of this condition as normal or even desirable (137).

Jim Dunlop (1992) articulated this when he wrote:

One of the saddest forms of alienation that currently exists...is the alienation of most people from the [E]arth itself and from the workings of natural systems whose operation is imperfectly understood....This alienation is chronic because education about the [E]arth and about the environment rarely entails more than the provision of facts about specific issues or problems. *To address alienation, environmental literacy beyond the curriculum is part of the challenge that confronts humankind.* What is needed is an increase in information which will allow people to filter the inordinate quantities of propaganda that pours out from government departments, environmental organizations and business, in order to make reasoned judgments. As a result, people would be introduced to a different value system which would perceive human-kind in a symbiotic relationship with the [E]arth. (80, emphasis supplied)

To illustrate, in 1993, I taught a General Ecology course for science majors at a private four-year liberal arts college (now a university). On the first day of class, after the normal administrative necessities, I asked the students to tell me why they had enrolled in the course (apart from the fact that it was required). Without exception, I was told it was because of the desire to "learn more about environmental issues, so we can help to save the environment." I asked if I could try something. I asked students to raise their hands if they could take me out to the campus nature trail and point out a beech tree (*Fagus grandifolia*) for me—a species common on the campus and rather difficult to confuse with any other species. Not one hand went up. I then said, "Okay, so you don't know trees. What about birds? How many of you could take me outside and tell me when you hear a robin (*Turdus migratorius*) singing?" Not one hand went up. This finding would probably not surprise Mike Weilbacher, who suggested that he could guarantee that not one in almost one million kids in Philadelphia-area schools can identify a grackle (*Quiscalus quiscula*) or recognize the song of the titmouse (*Parus bicolor*), both common in all Philadelphia-area habitats (Weilbacher 1993, 4).

Playing up the role, I acted frustrated and asked how many could take me to the mall and show me a CD player. Every hand was raised. "Okay," I said, "how many could, if I brought in a radio, recognize Whitney Houston singing 'I Will Always Love You'?" Every hand went up. I was flabbergasted! I said, "You mean to tell me you can't identify one of the most common trees on this campus or a birdsong you've heard all your life, but you can identify a human-made mechanical device that has only been around for about 10 years and a human song that has been around for even less than that?"

"But no one has ever taught us about those things (trees and birds) before" was the students' response.

Later that semester, I found myself teaching the water cycle to junior and senior science majors who acted as if they had never heard of it before! Biogeochemical cycles, energy flow, speciation, biodiversity, soil structure, and competition were topics new to these students—students who had already completed 15 years of formal, structured education! This finding would probably not surprise Larry Gigliotti, who suggested that "[w]e seem to have produced a citizenry that is emotionally charged but woefully lacking in basic ecological knowledge" (1990, 9). My experience was repeated twice while teaching an environmental studies course at a community college.

Ecological illiteracy is not unique to students. Robert Hazen reported that a Nobel laureate in chemistry told him he had never heard of plate tectonics (Pool 1991, 123). And Stephen Kellert reported that 73 percent of adults asked in a survey "were unfamiliar with the notion of the loss of biological diversity. The idea was new even to many of the 210 additional interviewees who belonged to environmental organizations" (Pennisi 1993, 410).

James Trefil stated, "If you don't know about something, you don't value it" (Pool 1991, 123). How, then, if students know so little about the natural world (the "environment," as so many call it), can they value it? If they don't value it, how serious can we expect their commitment to learning about it vis-à-vis "saving" it will be? What is it that they think they want to save? The point is that far too many people, educators and learners alike, don't really care enough about the "environment" to learn the basics of environmental literacy.

We teach children their ABCs and 1, 2, 3s because it is generally accepted that to be unable to read and write or to be unable to perform basic mathematical computations would put a child at a disadvantage. We value the ability to read, write, and perform basic mathematical computations. So critical, so valuable, are these skills considered that we ensure that they are components of children's formal education for 13 years. The point of this chapter is not to challenge that notion. Rather, it is to stimulate debate about what kind of competitive disadvantage we are at as a species if we lack environmental literacy and do not become what I call species-literate.

Contrast the knowledge of my hypothetical human who encountered ET with the following: Heinz and Maguire (n.d.) report that in Africa, a !Ko bushwoman is thought to have only average knowledge of local plant lore if she can identify and name 206 out of 211 plants in her domain in spite of the effects of a severe drought on the species' appearance. (That's like getting a letter grade of C after correctly answering 98 out of 100 questions on an exam!) Bushmen from the same domain have general botanical knowledge of at least 300 plants. Similarly, Conklin (1969) noted that in the Philippines, Hanunóo farmers can identify more than 450 animals and 1,600 plants! Of these 1,600 plants, 1,500 are considered "useful," and 430 of them are grown deliberately for their unique medicinal or other properties. Because they have such a refined classification scheme, the Hanunóo plant categories outnumber the taxonomic species classified by botanists by 400!

From Ecological Literacy to Species Literacy

All over the world, people in other cultures are able to identify their plant and animal brothers and sisters, classify their soils, predict the best days to plant or harvest crops based on phases of the moon, predict weather changes by observing cloud patterns, etc., to a much higher degree than many in the so-called educated cultures. Here in the United States, for example, I have been asked by supermarket checkout clerks to identify such items as spinach and nectarines! This was not always the case. For example, early European settlers of North America treated sore throats with yellowroot (*Xanthorhiza simplicissima*); this knowledge they no doubt obtained from Native Americans.

As another example of the failure of EE regarding ecological literacy, I have another personal experience to relate. In 1995, I accompanied my oldest son's kindergarten class on a field trip. The trip was to a local nature center. The primary purpose (I suppose) was to introduce the children to nature. Part of the trip involved a hike through an adjoining woodlot. On the trip, the naturalist (in this instance, I use the term with reservations) asked the children to stay alert for "monster brains." When she finally located one, she exclaimed, "Look, monster brains!" At this point, a tiny little girl let out a blood-curdling shriek and looked for a place to run. Realizing her mistake, the leader quickly tried to comfort her. What was to happen next both delighted me and also sensitized me to the problem that I have been focusing on in this chapter. My son, upon hearing "Look, monster brains!" challenged the leader by stating, "Ms. Jones (not her real name), those are not monster brains. They are from an osage-orange tree" (*Muclura pomifera*)—as my wife and I had taught him. I was left wondering if the little girl had ever been in the woods before and if she would likely return anytime soon!

In 1994, en route to the Campus Earth Summit at Yale University, I overheard two young female students talking about how concerned they were about the environment, how little they knew about it, and how much more they wished they knew. Finally, one turned to the other and said, "Look at that tree. I don't even know what kind it is." The tree was a sycamore (*Platanus occidentalis*)—as with the beech, hard to misidentify and very common. Yet even these students who admitted an interest in the environment did not possess this basic level of ecological literacy.

The type of education (which leads to what I call species literacy) is simply not currently available from most forms of traditional, formal EE. When I increase my students' awareness about tree identification, animal tracking, and plant lore, for example, they unanimously reveal that it is the type of education they want and thought they were going to get when they chose to major in an environment-related discipline. The alienation that currently exists between so many humans and the natural world is not only not acknowledged by many in traditional, formal EE circles; it is also trivialized by too many exercises whose potential to produce changes in learners' ecological literacy is questionable.

I am not trying to suggest that only nature study is relevant to ecological literacy. Rather, I am suggesting that it remains a missing element in the quest for ecological literacy. As an example, when my first son was about 3 years old, my wife and I began teaching him the letters of the alphabet. We were happy when he could simply identify each letter. Gradually, we expected him to reproduce the letters on his own. As he continued to develop, we expected him to combine two or three letters to make words. Combining words to make sentences followed. The point is that we recognized that there existed some foundational knowledge, skills, and abilities that had to be mastered prior to his advancing to each succeeding stage of development. We did not begin by asking our son to compose a 20-page, double-spaced essay when he was 3. For me, it is no mystery that he now voluntarily chooses activities associated with nature, including selecting what little TV he watches on the basis of whether the program involves some aspect of nature. Even now, in 1st grade, his classroom efforts center on nature. Even his lunch box is decorated with red-eyed tree frogs (*Agalychnis callidryas*)—his favorite—and other animals, while his classmates have Power Rangers,

Batman, Spiderman, and other such characters decorating theirs. This experience has been and is still being repeated with his younger brother, with the same results. Yet, in many instances, ordinary people are asked to perform such Herculean tasks on an everyday basis. How, for example, can people decide on paper versus plastic, cloth versus disposable, or how to vote on recovering the energy in discarded municipal resources if they do not possess some understanding of the laws of thermodynamics, and conservation of matter along with a value system that considers alienation from the environment as intolerable? How can people modify their behavior if they have no understanding of the ecological basis of human-environment issues? How can people who would be unable to survive if they were removed from their human-created surroundings embrace environmental preservation? So serious is the situation that some have chosen to make fun of it. For example, the Honda Corporation ridiculed the deplorable condition of ecological literacy in one of its commercials that aired in 1993. A customer in the checkout aisle of a grocery store is asked by a clerk, "Paper or plastic?" What followed were several rapid-fire images of the merits and demerits of each of the alternatives. The customer was presented as confused and unable to answer the question. However, even Honda hedges. In the final scene of the commercial, Jack Lemmon says, "Well, at least you'll know how to get your groceries home."

To modify what Trefil has suggested, I would say that you learn only about what you value. To learn about environmental preservation and the attendant behavior modifications necessary to accomplish such a goal implies a value system that cherishes the environment in a condition in which either human alterations are minimal or their impacts are ecologically benign. As Aldo Leopold once stated, "A thing is right when it tends to preserve the integrity, stability, and beauty of the biotic community. It is wrong when it tends otherwise" (Leopold 1949, 224—25). However, lacking ecological literacy, the average person has absolutely no understanding of what represents the "integrity and stability" of the biotic community. And if a person does have some understanding of the "beauty" of the biotic community, it is frequently framed only in the context of a human-created environment and not an undisturbed ecosystem. For example, I know many people who consider the skylines of Chicago and New York to be "beautiful," but the splendor of the Grand Tetons to be "scary" or "boring."

Conclusions

Recently, I was reviewing a videotape on the efforts to save the golden lion tamarin (*Leontopithecus rosalia*). At one point in the tape, the researcher involved with the effort discussed how a captive-bred male that was subsequently released back into its native habitat seemed to have learned how to survive "in the wild," a fact that caused her to be optimistic about this individual's potential to teach other captive-bred tamarins how to survive after their release.

This led me to think about the human condition, especially as it relates to the alienation mentioned earlier. Have modern humans become little more than an artifact of our ancestors—those who knew how to survive "in the wild"? Have we become more members of cultures and societies than a species on this planet? Are modern humans captive-bred to the extent that they need to be taught how to survive in situations outside of industrialized settings? If the masses are not receiving this kind of education, what can be said for the potential for long-term survival of *Homo sapiens?*

All is not lost! As more and more students become aware that there is an alternative, they are expressing an interest in programming that truly develops in them an understanding of environmental issues and environmentally responsible behaviors associated with ecological literacy. And, as the saying goes, create a demand and a supply will follow. Hopefully, a demand for instruction that provides indigenous knowledge to support ecological literacy will lead to increased opportunities for the field of EE, as well as improved opportunities for human societies to develop in ecologically sustainable ways.

References

Charles, L., J. Dodge, L. Millman & V. Stockley. 1981-82. "Where you at?" *Co-Evolution Quarterly* 32:1.

Comstock, A.B. 1994. *Handbook of Nature Study*. Ithaca, NY: Cornell University Press.

Conklin, H.C. 1969. "An ethnoecological approach to shifting agriculture." In Vayda (below), 229-30.

Dunlop, J. 1992. Lessons from environmental education in industrialized countries. In Schneider, Vinke & Weekes-Vagliani (below), 123-45.

Flicker, J. 1996. "Promoting a Lifetime of Learning." *Audubon 98* : 2, 6.

Gigliotti, L.M. 1990. "Environmental Education: What Went Wrong? What Can Be Done?" *Journal of Environmental Education 22* :1, 9-12.

Heinz, H.J. & B. Maquire. n.d. The Ethno-botany of the !Ko Bushmen: Their Ethno-botanical Knowledge and Plant Lore. *Occasional Paper No. 1*, Botswana Society. Gaborne, Botswana: Government Printer.

Leopold, A. 1949. *A Sand County Almanac.* New York: Oxford University Press.

Louv, R. 1990. "If People Pull Down Nature...." *Audubon, 75:* 5, 50-57, 106.

Orr, D.W. 1992. *Ecological Literacy: Education and the Transition to a Postmodern World.* Albany, NY: SUNY Press.

Pennisi, E. 1993. "What is Biodiversity, Anyway?" *Science News* (June 26), *143,* 410.

Pool, R. 1991. "Science Literacy: The Enemy is Us." *Science, 251,* 266-67.

Schneider, H., J. Vinke & W. Weekes-Vagliani, eds. 1992 *Environmental Education: An Approach to Sustainable Development.* Paris: Organization for Economic Co-Operation and Development.

Vayda, A. P., ed. 1969. *Environmental and Cultural Behavior: Ecological Studies in Cultural Anthropology.* Garden City, NY: The Natural History Press

Weilbacher, M. 1993. "The Renaissance of the Naturalist." *Journal of Environmental Education 25* :1, 4-7.

Chapter 27

Modeling the Greenhouse Effect at the High School Level

Patricia Carlson
New Trier High School
Winnetka, Illinois
and
Steven McGee
Northwestern University
Evanston, Illinois

A s classrooms shift from a textbook-lecture-lab model of science education (Ruopp et al. 1992) toward a model based on the practices of the science community, they will afford more opportunities for students who are traditionally under-represented in science. The textbook-lecture-lab model focuses on the abstract, conceptual qualities of science. A side effect of this approach is that some students come to view science as a static set of facts to be memorized but that have no relation to their life outside the classroom (Linn & Songer 1993).

A model of science education that focuses on the practices of the scientific community highlights the communicative, negotiated qualities and the personally meaningful qualities of science (McGee & Pea 1994). This chapter will describe an effort at New Trier High School (NTHS) in suburban Chicago to create a classroom environment based on the practices of the scientific community.

Environmental Science Studies Keyed to Student Interests

In recent years, NTHS has seen a dramatic increase of student interest in environmental issues. There are currently three environmental clubs at NTHS that engage in letter-writing campaigns, product boycotts, etc. The perception among the faculty is that students in these clubs react to environmental problems on a purely emotional level (Carlson 1993). The teachers feel that students are easily swayed by biased claims designed for an emotional rather than a scientific response. Once these claims become entrenched beliefs, they are often extremely difficult to change (Brewer & Chinn 1991).

Thus, Carlson, a former environmental chemist and currently a teacher at NTHS, created an environmental science course in the fall of 1992 to give students an opportunity to explore these issues from a scientific perspective. The goal of the course was to temper students' naive passion for environmental issues with enough scientific information that (1) they could make reasonable judgments as consumers, and (2) they would be able to critically interpret popular environmental literature. At the heart of the course is discussion of various viewpoints on particular issues, coupled with data-oriented explorations of these same issues.

During this same period, Carlson joined the Learning through Collaborative Visualization (CoVis) Project at Northwestern University. The CoVis project, in conjunction with the Department of Atmospheric Sciences at the University of Illinois, Urbana-Champaign and other academic and industry partners, offers a unique opportunity for students to explore phenomena that interest them, collect data, and, through discussion with other members of the community, try to understand what the data mean.

In the context of engaging in student-initiated science projects, CoVis students use a high-performance computing and communications network (wide-band ISDN) funded by NSF to gain access to atmospheric science data which will enable them to further their understanding of environmental science phenomena (Pea 1993; Pea & Gomez 1992). Through desktop video-conferencing, students can discuss their visualizations with members of the atmospheric science community.

In the fall of 1993, the students from the NTHS environmental science course conducted a six-week long project in atmospheric science, which is the basis for many issues in environmental science (e.g. global warming, acid rain, pollution). The current study examines student beliefs about science, attitudes towards their atmospheric science projects, and project artifacts.

Given the pragmatic constraints of today's classrooms, it is very difficult for students to engage in the practices of the scientific community. Although students might become interested in investigating atmospheric phenomena, they lack any of the necessary resources that the atmospheric science community routinely uses to study the atmosphere. Students normally do not have access to atmospheric science data and models. When they do have access, it is often in a form that is difficult to interpret or manipulate (Gordin & Pea, in press). Students also normally do not have access to members of the atmospheric science community. Therefore, in a traditional classroom, students do not have the opportunity to directly experience science nor do they receive indirect benefits by working with more advanced members of the community (Lave & Wenger 1991).

The CoVis project supports students' efforts to engage in the scientific community by providing access to the necessary resources. Through a high-performance computing and communications network, CoVis links teachers and students from two Chicago-suburban high schools (Evanston Township High School and New Trier High School), learning-science researchers from Northwestern University, and atmospheric science researchers from the University of Illinois, Urbana-Champaign. The network provides connections to the Internet and high-bandwidth channels for desktop video-conferencing and shared computer applications (e.g., for sharing and interpreting visualizations) among members of the network. In the context of both earth science and environmental science classes, students and teachers from both high schools have engaged in projects centered on weather and the environment, using data from the atmospheric science community.

Access to Atmospheric Data

The CoVis project has initially provided access to three sources of atmospheric data: the National Meteorological Center (NMC) Grid Point Dataset Version II, current National Weather Service (NWS) observations, and the Greenhouse Effect Detection Experiment (GEDEX). Each of these data sets is designed to be flexibly used by scientists in a variety of settings for a variety of research agendas. It is up to each scientist to customize the data for her/his own purposes. This may involve writing custom computer programs to reprocess the data or developing specialized templates and color palettes for visualization software (McGee & Pea 1994). In order to develop these customizations, a scientist needs to understand how the data were collected and how they are stored. Often, key information is not provided within the data itself since it is commonly understood by the community (Gordin & Pea, in press). This makes it quite difficult for students to use data from the atmospheric science community since they initially lack the skills to customize the data.

Working closely with atmospheric scientists from the University of Illinois, Urbana-Champaign, and from the University of Chicago, the CoVis project has developed customizations of the NMC data, the NWS data, and the GEDEX data that are specialized for the kinds of investigations that are conducted in the CoVis classrooms. The Climate Visualizer (see Gordin, Polman, & Pea, in press, for a full description) provides a customized student-appropriate interface with the NMC data. In particular, it provides a graphic interface with temperature, winds, and height contours at several levels of the atmosphere in the Northern Hemisphere. The data which range from 1962 to 1989, can be accessed as either daily readings or monthly means. Once a request is made, the Climate Visualizer displays the visualization of the data in a window, along with the supporting information. It provides labels for the variables that were used, latitude and longitude, etc. Students can use the Climate Visualizer for a variety of investigations. For example, students can subtract months from different seasons to investigate seasonal differences, or they can track historical severe-weather outbreaks.

The Weather Visualizer (see Fishman & D'Amico 1994 for a full description) provides access to the real-time satellite photos, weather maps, and station reports that the University of Illinois, Urbana-Champaign, generates from the NWS data. Most of the customization is accomplished at UIUC. The Weather Visualizer provides clear labeling of each image and Macintosh Balloon helps to explain some of the weather symbols that are used. Students can use the Weather Visualizer to make weather predictions or to track weather conditions over time.

Using the Radiation Budget Visualizer, students can gain access to GEDEX data to explore how incoming and outgoing radiation need to balance for the Earth to maintain a constant average temperature. Students can investigate levels of radiation, surface temperature, and the greenhouse effect by using three different models: no atmosphere, atmosphere but no clouds, and atmosphere with clouds. Options for visualizations are selected from a menu and are displayed in separate windows. This allows students to compare the different models with each other.

Data-centered Investigations

In order for students to conduct data-centered investigations, we need to do more than just provide access to data. There are several tasks that students need to be able to do in order to lead to a successful investigation using data. The assessment measures for the atmospheric science project made explicit the ways in which students were to use data. By directly assessing the desired behaviors (i.e., scientific practices), we created a systemically valid measure of project quality (Frederiksen & Collins 1989). The following paragraphs indicate some of the skills that are important for data-centered investigations:

First, students should know how to draw conclusions from data. Interpreting and drawing conclusions from data is a task that students are rarely asked to do in school. Students have difficulty exploring issues outside of textbook problems (Tinker 1992). Without methods for organizing data, students find it difficult to create theories about the data (Chinn & Brewer 1993). Providing students with alternative models can be useful if the models are (1) plausible, (2) of high quality, and (3) intelligible (Chinn & Brewer 1993).

Second, students should learn common quantities and magnitudes. Articles on environmental issues, whether in science journals or the popular press, routinely use the standard environmental science units. Some of these units measure quantities in terms that are difficult for students to imagine, such as ppb (parts per billion) or ppm (parts per million).

This usage has implications for public policy debates, such as the "Clean Air Act" or "Clean Water Act." Both pieces of legislation include reducing the concentration of certain contaminants to amounts measured in ppm or ppb or even requiring zero concentration. The cost of doing this could be astronomical, if at all possible. Students cannot truly grasp "how much" of some toxin exists without understanding these units.

Third, students should know how to identify acceptable error. The acquisition of data inevitably involves error. This is a crucial concept for science students to understand. The error inherent in the equipment used, error in the experimental design, and human error must be taken into account when ultimately trying to make sense of the data. For example, as a former environmental chemist, Carlson routinely performed limit of detection (LOD) tests on instruments so that her lab could qualify to analyze samples for the Environmental Protection Administration (EPA). Many of the concentrations of the test samples were actually at the LOD of the instruments. In other words, the instrument was barely capable of distinguishing the contaminant from background noise.

Scientists studying global warming must address issues of error. Given that there is some variation and uncertainty in historical temperature data, is a 0.5 degree Celsius average increase in global temperature statistically significant? Can the "signal" be distinguished from the "noise"? Is the world really getting warmer? Acknowledging the reality of error is particularly important when the data are used in calculations or massaged in some way (like a visualization) as that error is propagated.

Students are not sure how much of the data constitutes "evidence" or whether they are allowed to throw out anomalous data points that seem like outliers. Upon encountering data that do not fit their current belief, students and even scientists will ignore those data as "random error" (Chinn & Brewer 1993). For example, one group explored climatic temperature differences between the Pacific and Atlantic coasts of the United States. They gathered historical climate data by using the Climate Visualizer for four areas on each coast, and their conclusion was that the Pacific coast was generally warmer. Closer examination of their data, however, revealed that some of their data contradicted their conclusion. They simply ignored those data as random error.

Method

Students were provided with student-appropriate interfaces with atmospheric data and were explicitly assessed on their ability to use the data to conduct investigations. What other factors influence students' ability to engage in the practices of the scientific community? The current study correlates the beliefs of students about science, their attitudes toward their atmospheric science projects, and project artifacts. The quantitative analysis is supplemented with classroom observations by the teacher and the researcher and with selected student interviews.

The Setting

The NTHS environmental science course uses 40-minute periods with two double labs per week. The students have access to six CoVis workstations in the classroom. Each CoVis workstation is connected to the Internet and contains the suite of visualizers. Students have access to E-mail, newsgroups, and Appleshare for storage of files. There were seven project teams that were selected by the teacher. In creating the groups, the teacher tried to have a least one person in each group who might be able to play a leadership role. The teams were composed of either three or four students. Some of the projects used the Climate Visualizer ($n = 4$), and some used the Weather Visualizer ($n = 2$). One group used a weather almanac.

The Science Beliefs and Attitudes Survey

This survey explores students' beliefs about the nature of science. It was taken directly from Linn & Songer (1993). Students are rated as believing that science is a "dynamic" process that is constantly building upon previous work or believing that science is "static" and contained in the textbook. A third category is a mixture of the two. This instrument was administered early in the fall of 1993 to all of the CoVis students, along with a battery of other surveys for use in other CoVis research. The battery of surveys was administered again in late February 1994.

The Project Survey

This survey was administered individually to the team members immediately following the completion of their project. It explores students' attitudes toward the project (e.g., how much they felt they learned from the project, their satisfaction with the final product, the effort they put into it, and the project quality) and the cohesiveness of the project team. Part of this survey was adapted from Seashore's Group Cohesiveness Index (Seashore 1954), which measures how well the team worked together.

Student Interviews

The teacher and the researcher selected four students to be interviewed by the researcher. A boy and a girl were selected who the teacher felt enjoyed working with data and a boy and a girl who the teacher felt were not interested in working with data directly. The students were asked about their future plans, their reasons for taking the environmental science course, and their opinions about the projects they had conducted.

Student Profile

There were 25 students in the environmental science course who conducted atmospheric science projects. Table 1 shows the breakdown of the class. The course consisted of mostly seniors. Most students did not see themselves as pursuing a career in science. This is also evidenced by the fact that half the seniors were not taking physics, a required course for many science majors. Through structured interviews, we determined that there were some students who were not interested in a science career but wanted to learn about environmental issues and that some of the students who wanted to pursue a career in environmental science did not feel a need to take physics.

Table 1.

Breakdown of NTHS Environmental Science Class

Category	Students	
Total (n)	25	
Gender (%)	Male	46
	Female	54
Grade (%)	10th	4
	11th	8
	12th	88
Physics? (%)	Yes	43.5
	No	43.5
	Not senior	13
Science	Yes	26
Career? (%)	Maybe	19

Results and Discussion

Project quality

There were three measures of project quality: the project grade given by the teacher, each student's self-evaluation, and the average of each team's project evaluations. The project grade and the individual self-evaluation were significantly correlated, but the average team project evaluation was not correlated with the project grade. The students rated the projects roughly one letter grade higher than the teacher. Girls had a tendency to rate themselves lower than boys.

High-quality projects had team members who worked well together and who felt that they had learned from other students. Group cohesiveness and learning from other students were correlated. Learning from other students was significantly correlated with team members' helping each other, and helping each other was negatively correlated with learning from the teacher. Teams with a higher percentage of girls were more likely to help each other. The average science belief score of the project team was correlated with the group cohesiveness index.

Project teams from the top 50 percent of project grades were composed of students rated as either dynamic or mixed students. There were no static members on these project teams. Project teams from the bottom 50 percent were composed mostly of static and mixed students. In fact, there was one group that had no mixed or dynamic students. There was only one dynamic student in the bottom 50 percent of project grades. Research notes on these projects reveal that clear group leaders emerged for the upper 50 percent of projects and that there was relatively little leadership for the bottom 50 percent of projects.

Student characteristics

None of the characteristics considered so far were significant predictors of a student's science attitude. These include grade in school, gender, science course background, and desire to pursue a career in science.

An interesting pattern emerged among the students interviewed. Both of the students who were selected because the teacher felt they did not like to work with data were deeply interested in environmental issues. One student routinely participates in environmental activities (e.g. park cleanups, recycling programs). The other has changed his personal lifestyle since taking the course (e.g., taking shorter showers, encouraging his family to recycle).

Both students who were selected because the teacher felt they enjoyed working with data were not personally motivated to act upon any concern for the environment. Through an investigation into the environmental effects of paper versus plastic grocery bags, one student commented, "I learned that plastic isn't as bad as everyone has been saying." The other student commented, "I'm still going to drive a car even though it pollutes the environment. I'm not going to change my lifestyle."

Conclusions

Environmental science issues affect us all. By the time students enter a course like the NTHS environmental science course, they bring with them attitudes about environmental issues and beliefs about how to investigate them and how to act on their understanding of them. In a textbook-lecture-lab model of science education, these differences are masked. Since everyone is expected to learn the same material at the same time, there is no room for students to learn from the unique qualities that classmates bring to the learning situation.

In a classroom that is modeled on the practices of the science community, there is more of an opportunity for students to work with each other and benefit from the unique qualities of other students. In the current investigation, three out of the seven teams worked well together and learned from each other. Since these project teams contained members who held dynamic views of science, students were able to learn about the scientific process through their close interactions with other project team members. Unfortunately, the other four out of the seven project teams were not able to benefit from project team members who held dynamic beliefs about science. Future investigations will explore techniques for fostering success in all students.

Acknowledgments

This research has been supported in part by the National Science Foundation (#MDR-9253462) and our industrial partners Ameritech and Bellcore. We are grateful for hardware and/or software contributions by Aldus, Apple Computer, Farallon Computing, Sony Corporation, Spyglass, Inc., and Sun Microsystems.

We would like to thank the students of the NTHS environmental science course for their cooperation in this study. We are grateful to Douglas Gordin, who developed the Radiation Budget Visualizer and co-developed the Climate Visualizer with Joseph Polman, and Joey Gray and Barry Fishman, who developed the Weather Visualizer. We would like to thank other CoVis Project team members, including Roy Pea and Louis Gomez, the principal investigators, and also Daniel Edelson, Laura D'Amico and Kevin O'Neill, and Phoebe Peng. We are grateful to Mohan Ramamurthy,

John Kemp, Bill Chapman, and Steve Hall from UIUC and Ray Pierrehumbert from the University of Chicago for their contributions to the design of CoVis software.

References

Birnbaum, L., ed. 1991. *The International Conference on the Learning Sciences.* Evanston, IL: Association for the Advancement of Computing in Education.

Brewer, W.F. & C.A. Chinn. 1991. "Entrenched Beliefs, Inconsistent Information, and Knowledge Change." In Birnbaum (above), 67-73.

Carlson, P. 1993. "Environmental Investigations: The Science Behind the Headlines." *The Science Teacher 60,* 34-37.

Chinn, C. A. & W.F. Brewer. 1993. "The Role of Anomalous Data in Knowledge Acquisition: A Theoretical Framework and Implications for Science Instruction." *Review of Educational Research 63:* 1, 1-49.

Fishman, B. & L. D'Amico. 1994. "Which Way Will the Wind Blow? Networked Computer Tools for Studying the Weather." Paper presented at the meeting of ED-MEDIA 94, Vancouver, BC, Canada.

Frederiksen, J. R. & A. Collins. 1989. "A Systems Approach to Educational Testing." *Educational Researcher 18:* 9, 27-32.

Gordin, D. N., & R.D. Pea (submitted). "Prospects for Scientific Visualization as an Educational Technology." *Journal of the Learning Sciences.*

Gordin, D. N., J. Polman & R.D. Pea. (in press). "The Climate Visualizer: Sense-making Through Scientific Visualization." *Journal of Science Education and Technology.*

Lave, J. & E. Wenger. 1991. *Situated Learning: Legitimate Peripheral Participation.* New York: Cambridge University Press.

Linn, M. & N.B. Songer. 1993. "How Do Students Make Sense of Science?" *Merrill Palmer Quarterly 39:*1, 47-73.

McGee, S. & R. D. Pea. 1994. "Cyclone in the Classroom: Bringing the Atmospheric Science Community into the High School." In *Proceedings of the 74th Annual Meeting of the American Meteorological Society.* Nashville, TN: American Meteorological Society, 23-26.

Pea, R. D. 1993. "Distributed Multimedia Learning Environments: The

Collaborative Visualization Project." *Communications of the ACM, 36:5,* 60-63.

Pea, R. D., & L. Gomez. 1992. "Learning through Collaborative Visualization: Shared Technology Learning Environments for Science." *Proceedings of SPIE '92: Enabling Technologies for High-Bandwidth Applications, 1785,* 253-64.

Ruopp, R., S. Gal, B. Drayton & M. Pfister. 1992. *LabNet: Toward a Community of Practice.* Hillsdale, NJ: Lawrence Erlbaum.

Seashore, S. 1954. *Group Cohesiveness in the Industrial Workplace.* Ann Arbor, MI: University of Michigan.

Tinker, R. F. 1992. *Thinking About Science.* CEEB.

Chapter 28

Vehicles: Metaphors for Environmental Education

Martin Stanisstreet
and
Edward Boyes
University of Liverpool
United Kingdom

E ven nonalarmist sources predict an increasing impact of environmental problems on society. In the near future, citizens may need to make individual decisions about their lifestyle, and societies may need to impose restrictive legislation to reduce further degradation of the environment. The efficacy of such personal decisions and the acceptability of such legislation may be influenced by an appreciation of why they are necessary.

One such lifestyle pattern is concerned with car use. Cars have brought improvements in our quality of life in terms of convenience, recreation, and extended possibilities of travel, but, at the same time, the extended use of vehicles is increasingly a cause of environmental degradation on a local, national, and global scale. It is important, therefore, that children be informed about how car use may impact on the environment.

We have started to try to unravel the mental models that students use in their thinking about cars and the environment because an understanding of how students are thinking may guide teaching strategies. The teaching theme of "cars" may be a useful entry into helping students clarify their thinking. Such a topic combines students' natural enthusiasm for matters environmental, their intrinsic interest in cars, and the common perception of vehicles as polluters.

"I don't like the smoke but I like riding in the car" (Stephen, age 7)

Stephen encapsulates the global dilemma of the car. In the industrially developed world, cars are an integral component of the environment. But, at the same time, the expanded use of cars is increasingly a cause of environmental degradation. Vehicle manufacture and use deplete resources, some of which are nonrenewable. Construction of roads, bridges, and parking spaces demands land. Cars use results in noise (see Bronzaft, this volume), physical harm to individuals in traffic accidents, and noxious exhaust emissions. These emissions have effects on human health, other biological repercussions, and physical consequences on a local, national, and global scale. When vehicles have finished their useful life, they require disposal of unsightly and potentially hazardous materials.

For their part, car manufacturers have made efforts to reduce the adverse effects of their products on the environment. More fuel-efficient engines and post-combustion devices, such as catalytic converters, have been developed. There is a move toward making a higher proportion of car components from materials that are recyclable when the vehicle is discarded. Despite these efforts, it is anticipated that the car will continue to have an impact on the environment because the increasing demand for cars worldwide and the greater mileage covered by each car will outstrip the advances in technology that may reduce the impact of individual vehicles. Clearly, then, the future decisions of today's children regarding the ownership and use of vehicles will have considerable environmental implication. In light of this fact, it is important that they be well informed about the issues and the consequences of their decisions.

Students' Alternative Conceptual Frameworks

Research in science education has suggested that students attempt to make sense of the information they receive; they try to construct individual ideas into a conceptual framework (Driver 1983). Unfortunately, this conceptual scaffold that "holds" the individual ideas is not always congruent with that of the accepted scientific understanding. Furthermore, because these frameworks have been constructed by individual students and are therefore "owned" by them and because they have been constructed to fit many other ideas assumed to be correct, the frameworks are robust and resistant to modification (Ausubel 1963). According to this view, teaching becomes not only a matter of supplying new information but also a matter of helping students to challenge their own conceptual frameworks and, if necessary, modify them to more orthodox frameworks (Driver et al. 1994). For effective teaching, therefore, certain questions arise. What are the alternative concepts held by students in the subject area to be taught? Why did these conceptions arise? How can they best be supplanted?

When we first became interested in this area, a review of the literature showed that there were many resources for EE: suggestions for schemes of work, courses of study, and so on. However, and surprisingly in the light of the importance of EE, there appeared to be much less primary research to provide systematic information on students' notions about environmental issues. The aims of our research program are to explore students' ideas about environmental issues and to make the results available to educators so that their educational strategies and resources can be most effectively targeted to meet students' needs. Here we report observations of students' ideas about the environmental impact of motor vehicles, the results of studies into how these ideas differ in students of different ages, and an exploration of the factors underlying students' ideas.

Our first step was to gather students' ideas about how cars might affect the environment. We used an open-form questionnaire containing four questions about how vehicles might harm the environment ("the world around us") and people and what might be done by vehicle manufacturers and owners or drivers to minimize the environmental impact of cars. In total, over 700 students spread through grades 6 (11-12 years), 8 (13-14 years), and 10 (15-16 years) helped in this study. All written responses were scrutinized to construct a list of "concept categories." The responses of each student were then reexamined and coded according to whether a particular category had been expressed or not, and these data were entered onto a computer datafile.

Students Think in Different Ways

The students appeared to think in different ways about the social and environmental costs of motoring. *First,* students were aware of the danger of direct physical injury from vehicles. *Second,* many students raised ideas of vehicle emissions, air pollution, and human respiratory problems. The ideas in this second batch were raised more frequently by older students, suggesting a development that embraces the more abstract, indirect effects of vehicles. *Third,* students, especially older students, raised ideas about global environmental problems. However, the problem with which vehicles were most frequently associated, erroneously, was damage to the ozone layer. We suggest that this persistent misconception is a product of a more general failure to distinguish between specific causes and consequences of particular environmental problems.

Most students thought that there was some potential for manufacturers to reduce the environmental impact of vehicles, although some responses expressed rather general ideas, such as that manufacturers should "reduce pollution" or "reduce emissions." There may be a danger here in students thinking that manufacturers have the technology to reduce vehicle emissions and that all that is required is for manufacturers to apply this technology. In turn, this might lead to an abdication of individual responsibility, with students thinking that the problem could, and should, be solved by "someone else," in this case vehicle manufacturers. The major specific idea, raised by about a quarter of the students overall and growing in predominance in the older students, concerned the manufacture of cars that use unleaded gasoline. This idea, the use of unleaded gasoline, was also the idea most frequently raised in response to the final question, on what owners or drivers could do to reduce the environmental impact of vehicles. Although unleaded gasoline has indeed made a contribution to the reduction in one health hazard of vehicles (that of lead poisoning by chronic accumulation of lead in the body), we again suspect that there may be a problem here. Few students raised the idea of car emissions resulting in mental impairment, so it may be that students are not aware of the specific problems wrought by leaded gasoline but just have an image of "environmental friendliness" of unleaded fuel.

In other studies, we have shown that students think, erroneously, that the use of unleaded fuel will reduce the effects of vehicles on global warming (Boyes & Stanisstreet 1993). There may even be a double misconception in the minds of some students, that unleaded fuel will reduce the "effect" (in reality, nonexistent) of cars on the ozone layer.

These ideas, together with an over-optimistic view of what catalytic converters can achieve, might persuade students that the use of unleaded fuel and the fitting of a catalytic converter make a car harmless to the environment and that no other action need be taken. In answer to the question concerning what drivers could do to minimize the environmental impact of cars, nearly one fifth thought there was nothing drivers could do. Also, few students suggested actions that could be taken independently of the type of car owned, such as nonaggressive driving and good engine maintenance.

Predominance of Ideas in Students of Different Ages

The next step was to examine the preponderance of individual ideas in students of different ages, to gain insight into the extent to which alternative ideas persisted in older students and so probably in the adult population. For this, we conducted a cross-age study using a pre-coded questionnaire consisting of a number of statements to which students responded by checking boxes labeled "I am sure this is right," "I think this is right," "I don't know about this," "I think this is wrong," and "I am sure this is wrong." A total of nearly 800 students spread through grades 5 (10-11 years), 6 (11-12 years), 8 (13-14 years), and 10 (15-16 years) completed the questionnaire. The questionnaire consisted of three sections, with questions about the consequences of increased vehicle emissions, the components of vehicle emissions, and the ways in which the environmental impact of vehicle emissions might be reduced.

Most students appreciated the consequences of increased vehicle emissions on a local basis, in the production of smog; on an international dimension, in the formation of acid rain; and on a global scale, in the exacerbation of global warming by the greenhouse effect. However, as our previous study had suggested, the connection drawn by students between vehicle exhausts and environmental problems was somewhat indiscriminate, in that more than three quarters of the students thought that vehicle emissions damaged the ozone layer. Furthermore, this misconception did not diminish much in the older students, suggesting that it might persist in the adult population.

A similar general picture emerged from the responses to statements concerning the biological consequences of increased vehicle exhausts. Most of the students appreciated that vehicle emissions are connected with respiratory problems, but some also, erroneously, associated vehicle exhausts with rainforest destruction or species extinction. Even in the oldest group of students, about one quarter held these views. As previously, we suspect that these views are a reflection of a more general, underlying lack of distinction between the different causes and consequences of different environmental problems.

Students seemed reasonably well informed about the components of vehicle exhaust emission. Most of the oldest group realized that car exhausts contain carbon monoxide (CO), carbon dioxide (CO_2), and water vapor (H_2O), although the last two ideas were more prevalent in the older age groups studied. Nitrogen oxide gases (NO_{xs}) were less well known, as was sulfur dioxide (SO_2), despite the fact that the nitrogen oxides from vehicles contribute to the formation of acid rain, which was well known by students. Most of the oldest group of students appreciated that ozone (O_3), oxygen (O_2), radioactive gases, and chlorofluorocarbons (CFCs) are not components of exhausts. The last observation is of interest because other studies have shown that many students know of CFCs as the cause of ozone layer degradation (Boyes & Stanisstreet 1994; Chambers, Boyes & Stanisstreet 1994), yet many in the present study thought cars were responsible for producing ozone layer damage.

The final section of this questionnaire explored students' ideas about how the environmental impact of cars might be reduced. The most prevalent idea, held by more than three quarters of the students, was that battery-powered vehicles could be used. In this context, it would be useful to explore the extent to which students appreciate that the generation of energy to recharge such vehicles, albeit at a site remote from that of vehicle use, does have an environmental cost. Only about half (even of the oldest students) recognized the correlation between gasoline consumption and pollutant emission or realized that vehicle maintenance was important, although the former of these ideas appeared to diminish in older students.

Why Do Students Hold These Ideas?

Most recently, we have completed a study of some 1,600 students in grade 9 (14-15 years) to explore the reasoning underlying their views of the links between vehicles and three environmental problems, global warming, acid rain, and ozone layer depletion. The questionnaire used was of a graphic nature, with arrows representing causal links between ideas. On each arrow were boxes labeled "Yes," "Don't know," and "No"; students were asked to check one box. The questionnaire was arranged to test possible specific "routes" of reasoning. Thus, it explored whether students thought that car exhausts contained certain components (carbon dioxide, CFCs, sulfur dioxide, nitrogen oxides, water vapor, acid gases, and heat) and whether each of these gases contributed to the three environmental problems (global warming, acid rain, and ozone layer damage). It thereby revealed the ways in which students might think that cars damage the environment.

Most students realized that car exhausts made global warming worse, and it is the reasoning of these students that we consider next. For example, the most popular reasoning "route" of these students appeared to be that of carbon dioxide. More than half appreciated both that car exhausts contained carbon dioxide and that carbon dioxide exacerbated global warming. Although three quarters of the students thought that CFCs are greenhouse gases, only one quarter thought (wrongly) that they are a component of car exhausts, so few thought that cars made global warming worse by this mechanism. About a quarter of the students realized that car exhausts contain oxides of nitrogen, and a third appreciated that these gases contribute to global warming. However, only one tenth of the students held both of these ideas, so only this proportion might realize that car exhausts exacerbated the greenhouse effect in this way.

About half the students realized that cars caused acid rain. In these students, the most popular "route" appeared to be the idea of carbon dioxide; nearly half of these students realized that car exhausts contain carbon dioxide and also thought that carbon dioxide caused acid rain. One quarter of the students thought that cars emitted sulfur dioxide and appreciated that sulfur dioxide contributes to acid rain. Such students might well be misinformed about the role of vehicle emissions in acid rain by this route, for although sulfur dioxide does make a major contribution to acid rain, the contribution to the atmospheric load of sulfur dioxide from cars is minor. Finally, more than three quarters of the students thought that car emissions damaged the ozone layer. Within this group of students, the most common conceptual "route" was again apparently via carbon dioxide; nearly half of this subset of students both realized that car exhausts contain carbon dioxide and thought that carbon dioxide damaged the ozone layer. About one quarter of the students thought that the heat from car exhausts caused ozone layer depletion.

Questionnaire Repeated with Second-Year Students

We have also repeated this questionnaire with a group of 100 second-year undergraduate students taking an Ecology course. Although this group of undergraduates had not been taught specifically about the issues studied here, we might expect them to be concerned and informed about environmental matters. Despite this, we found that some of the undergraduates retained misconceptions. For example, about one third thought that cars were responsible for ozone layer depletion, although some of these offered no "route" by which this might happen. The fact that this group retained some misconceptions leads us to think that such ideas are likely to be present in a proportion of the adult population.

Conclusions

As we explored students' ideas, a number of interlocking reasons for the conceptual difficulty of global environmental issues occurred to us.

First, ironically, in view of the fact that these issues are of global scale, these problems are imperceptible to individuals. The exhaust gases from most vehicles are invisible, and intangible emissions might not be readily appreciated as pollutants (Brody 1992). Furthermore, we cannot sense global warming, the acidity of acid rain, or ozone layer depletion. Thus, students' learning about these ideas cannot be experiential.

Second, these problems are associated with a degree of uncertainty. Experts do not agree on the extent of the problems, so students may receive confusing messages. Environmental lobby groups may be tempted to deny uncertainty or even to exaggerate the problems, whereas those with an interest in short-term sociopolitical stability may minimize them.

Third, (and this reason is partly connected with the uncertainty), many of the phenomena and their predicted consequences are stochastic events. Thus, a comprehension of risk and chance is required, and the notion of probability is known to be difficult for students. Next, these environmental problems are long-term, chronic problems and, as such, are more difficult to appreciate than acute, catastrophic events. Furthermore, young students in particular have poor "foresight" and prefer short-term pleasure despite long-term risks. Finally, these issues are multidisciplinary and require an understanding of a range of traditional "curriculum" sciences at a quite complex level.

In addition, the high profile of environmental issues, the fact that they are featured often in the popular media, may itself cause problems because students will gain much of their information from out-of-school sources. Even if the information given by these sources is accurate and balanced, there is no opportunity for students to interact with colleagues or peers to determine that they have not misunderstood the information. Furthermore, is there a danger that environmental problems suffer from over-exposure, that students develop "environmental fatigue" in the same way that people, faced with constant requests for support for good causes, develop "charity fatigue"?

We are faced with the prospect of education about important but complex and difficult ideas. Is education about environmental issues a lost cause? The students themselves give us more optimism. Students have a natural enthusiasm and concern for the environment. They do not need persuading that environmental issues are important or that the science of environmental issues is "relevant." Similarly, cars appear to appeal to students, and most students will have direct, everyday experience of vehicles. Linking these two subjects, cars and the environment, might be a useful way in which to start to address some of the alternative ideas revealed by our studies, especially since students see humanmade artifacts (Ali 1991), such as cars, as pollution sources.

Furthermore, some of the instruments we have used in our study, particularly the graphic questionnaire, which explores possible mental reasoning "routes," might be useful teaching tools in helping students challenge their conceptual frameworks. Exploration of the idea that exhaust emissions are a cocktail of different pollutants, already established in principle in students, might persuade them to start to dissect their rather general model of pollution, in which all sources of pollution result in all consequences of pollution, and to think in terms of a series of separate pollutants. This might lead, with older students, to a discussion of the reactions of a primary pollutant that generates secondary pollutants.

As students clarify their thinking about the consequences of an unrestricted expansion of vehicle use, they may be encouraged to consider their own future use of cars and, indeed, wider forms of "environmentally friendly" behavior. Perhaps the problem contains the seed of part of the solution; perhaps vehicles can be used as metaphors of EE.

Acknowledgments

We wish to thank our colleagues Debbie Batterham, Matthew Hillman, and Emma Leeson, who contributed to this project; the teachers who gave of their time to assist us; the students who answered the questionnaires; and the Ford of Britain Trust and Vauxhall Motors Ltd. for financial support.

References

Ali, I.M. 1991. "How Do English Pupils Understand Pollution?" *Environmental Education and Information 10,* 203-20.

Ausubel, D. 1963. *Educational Psychology: A Cognitive View.* New York: Holt, Rinehardt & Winston.

Boyes, E. & M. Stanisstreet. 1993. "The 'Greenhouse Effect:' Children's Perceptions of Causes, Consequences, and Cures." *International Journal of Science Education 15,* 531-52.

Boyes, E. & M. Stanisstreet. 1994. "The Ideas of Secondary School Children Concerning Ozone Layer Damage." *Global Environmental Change 4,* 317-30.

Brody, M.J. 1992. "Student Science Knowledge Related to Ecological Crises." Paper presented to the American Educational Research Association (AERA), San Francisco.

Chambers, W., E. Boyes & M. Stanisstreet. 1994. "Trainee Primary Teachers' Ideas about the Ozone Layer." *Environmental Education Research* 1:2.

Driver, R., A. Squires, P. Rushworth & V. Wood-Robinson. 1994. *Making Sense of Secondary Science: Research in Children's Ideas.* London & New York: Routledge.

Driver, R. 1983. *The Pupil as Scientist?* Milton Keynes, UK: Open University Press.

Chapter 29

Using the Econet to Promote Effective Environmental Education for Learning-Disabled Youth

Esther Zager Levine
Valley Stream North High School
Franklin Square, New York

This writer has long used computers in the resource room classroom with learning-disabled junior and senior high school students. Computers free these youngsters from their disability (especially in areas of handwriting, spelling, and written expression) and enable them to participate in classes on the same level as their nondisabled classmates. Despite these gains, learning-disabled youngsters continue to fall behind their peers in science. To increase the possibility for their success, the author has combined computer literacy with a specific telecommunications network. Telecommunications is a way of connecting real-life events and experiences to the classroom.

After extensive research into telecommunications projects, the writer determined that Econet, an ecology and environmental network, most closely met the needs of the students. Econet is an international computer-based communication system committed to serving organizations and individuals who are working for environmental preservation and sustainability. Learning disabled students need much exposure and experience to real-life science and technology projects in order to give them the background needed for understanding their courses and for making educated choices about future coursework and careers. Most environmental education organizations are part of the Econet network as are the Global Laboratory and the Kids '93 projects sponsored by National Geographic. The network allows users to dial locally, access information resources, use E-mail, and participate in public and private conferences.

The Research Problem

This project, specifically using the Econet network, was expected to affect the science education of learning-disabled youth in four important ways. (1) Students would have current information available on environmental and ecological concerns. (2) Students would participate in technology, such as E-mail and teleconferencing, and would be part of a team or study group. (3) Students would gain familiarity with the work of those in fields related to the environment and become familiar with the study of ecological problems. The network would provide students with information about the world of work in science- and technology-related fields. (4) Finally, the student, while gaining confidence and knowledge, would become a participant in his or her regular science class and would seek other science experiences in the future. It was further anticipated that the writer and her students would serve as a resource to teachers and others and would be able to model and demonstrate the usefulness of this technology. All of these expectations were met by the end of one school year of instruction in this telecommunications project.

Prior to Implementation

Two directors were approached, the computer director for the district and the building chairperson of science, to discuss the project. To answer the concerns raised at the meeting with these two directors, this author wrote to a dozen people who were already online with Econet and asked several questions of each person. In preparation for the project, this author also wrote to several organizations for grant money to cover the cost of the telephone during the project.

Response from the people and organizations contacted regarding the network was most positive. All of the network users contacted called or wrote and were informative about the network and what could be done on it. The directors were much more favorable toward the project, since all the information received was shared with them both.

The school principal was informed of the project. He seemed pleased and was positive and encouraging. The Director of Computer Use, who agreed to become the verifier for the practicum, felt that funding would be provided through the District's Teacher Center's mini-grant program. It eventually was. No outside grants were obtained to finance the work.

Conversations with students and teachers were completed, and surveys were given to all of the science teachers. They revealed some negative attitudes toward methods needed to teach learning-disabled students and accommodations necessary for testing. For example, one teacher was not using cooperative learning activities in his Biology and Life Science classes because he found them too time-consuming. A teacher of physical science didn't think the learning-disabled kids could answer the global thinking project question about carbon dioxide management because he "never taught them that"! And an Earth Science teacher was convinced that resource room students do not belong in Regents-level classes.

It was hoped that as the students showed greater interest in the science classes, their teachers would become more positive toward their abilities in those classes. For a variety of reasons, this expectation was the least met of all those in the project.

Actions Taken in One School Year

September

Students were oriented to the concept of the network chosen for the study. The importance of ecology and environmental studies in everyday lives, as well as in the future, was discussed. A bulletin board depicting telecommunications was prepared for the classroom before the students entered school.

The author interviewed the students regarding environmental issues. In the classroom, theme scrapbooks were created from magazines on global and environmental issues of the student's own choice. Topics that were depicted in this way were oceans and water, air pollution, waste management, and global issues, such as AIDS, war, and famine. Three books were created that focused on people and their problems, animals, and environmental issues concerning air, land, and water. The books contained slogans, collages, poems, and pictures.

The concept of telecommunications was directly taught. Lessons and discussion included understanding principles of telecommunications and understanding E-mail.

October

Classroom cooperative learning groups were formed among students who were in similar science courses, had similar interests, or simply were in the same resource room period. Students met and decided on one or more topics to research via online information sources.

Classes began E-mail and online searches for data related to the topics of interest chosen by each group. First, students were given a listing of the organizations and users currently online in their area of interest. E-mail was composed and then sent to several organizations. It was difficult to get students to draft inquiries before going online, but once they understood the process, they were even more reluctant to use the computer themselves to send E-mail. They were all reluctant to spell and write. They were most uncomfortable in the math-lab environment (in which the project was housed) and preferred to watch as the information they wrote or composed was input by the author. The interest level was extremely high, and the students were excited about the project. As one 8th-grader said, "This is fabulous. It is the way of the future." One of the early disappointments was that none of the E-mail was answered! We later learned that scientists like to answer very specific and direct questions, not general information ones, and that answers would best be obtained through conferences. Since we were learning the process of telecommunications together, the students and the writer were not easily discouraged.

By the end of October, some teacher interest was beginning. The chair of science visited one of the groups while they were telecommunicating, and he noted their excitement. The math-lab assistant became extremely interested in learning about it. The other resource room teacher was interested in seeing the project in action. The verifier for the project (who is the district-wide person in charge of computer applications) continued to encourage the search for grant money to pay for phone costs.

November

Work on the network continued each week. E-mail continued for the first and second week of the month, and then the concept of conferencing on the network was introduced. In addition to online information, printouts of conference materials taken from Econet at the author's home were brought into the resource room classroom. This was added to the materials being read from various newspapers and magazines. One popular piece from the Environmental Announcements Conference was called "Rad Day Ozone Alert"! In this communication, students learned about a boycotting action that was taken by a college group in the state of Washington to bring about public awareness of ozone-depleting products that are still on the market.

This became a topic of discussion for several days.

Late November also brought funding in the form of a mini-grant from our Teacher Center for the cost of the telephone time for telecommunications. The grant also provided a vehicle for dissemination of the project through our district's "Lights On for Education" exhibit, which occurred in March.

Students continued to explore topics on the environment and global issues through the almost weekly discussions of news items found on the network, along with news from newspaper and magazine articles brought to class for this purpose.

December

The work on the network provided us with many topics for reading, writing, and oral discussions. The students seemed to enjoy the lessons focused on these topics.

The three weeks of this month were spent working on the network in two ways: (1) by using conferencing for gathering information on topics of interest, and (2) by participating in the Kids '93 project. A group of youngsters age 15 or under began registering for pen pals on the Kids '93 project. To do this, they had to answer four questions about themselves, their attitudes towards global and local environmental problems, and their wishes for the future. They then had the author post their answers on the Kids' Cafe Conference for a response. Many of the students began composing letters to post to this online source. This was a most exciting part of the project for the students and was written about in the newsletters the author published for the community.

On two successive Fridays, videos depicting science and global information were presented in the classroom.

January

Students discussed topics of interest to be used for science fair projects. Because of delays and holidays, we had to review the three concepts we had already covered: sending E-mail, contacting conferences, and searching archives of conferences. At the same time, some of the younger students began to access the Kids '93 project for responses to their queries.

January 14 was the day of the first field trip of the year planned exclusively for resource room students. The author chose to take the students to

Brookhaven National Science Labs, where the first nuclear reactor was built and used, and to Cedar Creek Water Management Plant, where our own wastewater is cleaned. This trip was intended to acquaint students with some of the science-related enterprises in their own backyard. Student questionnaires following the field trip revealed that most of the students enjoyed the hands-on exhibits at Brookhaven Labs. "This was cool in a way because there was stuff to do that was fun," said one. But they would have liked more touring of the actual buildings. Although their comments about Cedar Creek varied greatly, they were mostly disappointed that there was too little touring, because of construction on the site, and too much lecturing. "Sitting and watching the film and lecture was boring. It would have been better if we could have been in the buildings," one student said. Once again, this reinforced the fact that learning-disabled students need to see and do things and not be lectured to.

A major source of frustration was having the telecommunications equipment in another part of the building. Throughout the year, grant moneys were sought to move the telecommunications equipment setup into the resource room classroom. When one group realized the possibility of having the equipment in our own classroom, the 8th-grader in the group turned to the older students and said, "I'm so lucky to just be getting into the resource room now. It's so exciting!"

February

During the month of February, there is usually a science fair at the school. This year, however, it consisted of the honors (gifted) students only. It was anticipated that some resource room students or teams would help gather data on projects accepted this year. However, although teachers were told of the availability of the students and the network, none of them took advantage of the help. Of the 90 projects in this year's science fair, only 15 pertained to issues of environment, energy, or ecology. This is less than 20 percent.

Non-honors students started on science projects, reports, and experiments for either March or April dates in science classes. The number of ecology and environmental projects chosen was very gratifying. Over 70 percent of the students who were doing reports and projects chose ecology, environment, or energy-related ones. We searched the archives and other sources for information on the following topics: recycling, solar energy, weather and the greenhouse effect, alternate power from wind- and solar-power sources, and electromagnetic fields and their possible dangers. The student projects were well done and well received. The average grade was above 80, with many students receiving 90s and 95s on their total projects.

During February, some networking began among these students and other participating student groups. For example, the Kids '93 project began to take off. There were letters from students in Denmark for several students and a contact from Huntsville, Alabama. Other students wrote to students in the United States and to other "kids" in foreign countries. Everything was done to locate groups that were participating in the Kids '93 project, and the younger students were all encouraged to continue reaching out on the Kids Cafe Conference. Through the Kids '93 project, collaborations were encouraged. It was exciting to the students whenever the news that someone had received E-mail was posted. This became the main topic for the second public-information newsletter published by the author.

Whenever teacher interest appeared, students were asked to volunteer to explain the session for the teacher. Those teachers who did witness the project were always impressed by the students' ability to explain what was going on.

March

As part of the funding received from the Teacher Center, the writer participated in "Lights On for Education" on March 4, 1993. This is a district forum for exemplary programs sponsored by the PTAs of the four districts in the community. Student cooperative groups were encouraged to help put together materials for the "Lights On for Education" displays and demonstrations. Many parents, legislators, businesspeople, and others from the overall community attended. Students explained E-mail, conferences, and data searches of the archives through information that was included in two newsletters issued to the public. Special boards designed to demonstrate the Econet telecommunications project were made. One student designed the background board for the display, and two newsletters that had been produced by the author were distributed to parents and community members. Students were excited to see the "Resource Room Newsletter No. 2," where their work was quoted. Hard copies of correspondence, E-mail, journal entries, etc., were also put on display to help explain the project to the public. Much favorable publicity resulted from the "Lights On" presentation, among which was news coverage in the Teacher Center paper, as well as local community papers.

On the basis of his familiarity with the project from the "Lights On" presentation, the school principal agreed to recommend bringing the designated phone line into the classroom. This was accomplished on March 25. The cost to the district was $247.30, which was authorized by the Business Office.

On March 31, the resource room students used the network in the resource room for the first time! The student response was wonderful. Three science teachers also watched one session. The level of enthusiasm about the project, from the author, classroom teachers, and the students, was at an all-time high!

In addition to demonstrating the project to any interested teachers in this school, offers have been made to make a presentation to all district teachers at the Teacher Resource Center next fall. Students will be asked to participate in the presentation.

April

April 22 was Earth Day. No student group at this school had participated in it before this group. We had a "slogan" contest among the various resource room classes. The student council sponsored a "pizza party" for the winners, and the sole judge of the slogan contest was a member of the Science Department. The winning slogan said, "The world is green, so lets keep it green. The sky is blue, so let it shine through." The slogan was later put on a poster by the students and is on continuous display in the main entrance corridor of the school. The poster displays a globe of the world, surrounded by blue on one side and green on the other, with the slogan across the entire poster. It is very eye-catching.

May

The final week of the project was the week of May 6. The post-survey was administered to all students in the resource room, despite the fact that some had much more exposure to the network than others. This was done because all of the resource room classes were exposed to films, articles, and discussions about ecology and the environment regardless of how much time they were able to spend on the network. Results from the pre- and post-survey are briefly summarized and discussed below.

Survey Results

Survey results were compiled and compared for each grade and for male and female students. The survey was scored + (positive) or — (negative) in regard to a given item, and that score was summarized in the form of a mean score for each student, classified by grade and sex, and then summarized again at the end of the project. The difference in the scores was the basis for the analysis of the results as more positive or less positive.

Junior high school learning-disabled students in grades 7, 8, and 9 became more positive with regard to their attitudes on world issues, ecology, and resources and have moved from negative to positive attitudes toward accepting positive moral values. They also became more positive toward themselves and less positive toward their school.

Survey results showed that senior high school students became more positive toward world issues and ecology, as well as accepting positive moral values, with the exception of the 12th-graders, who changed very little. The 10th-and 11th-graders became more positive in their attitudes toward themselves and less positive toward their school.

With respect to gender differences, in the pre-survey, it appeared that female learning-disabled students were slightly more positive than male learning-disabled students; however, it was clear that self-concept was low in both groups. The post-test results for all students showed a gain of more positive attitudes toward all of the variables except their attitudes toward school. On the whole, female students surveyed were more positive on world issues and themselves, while male learning-disabled students were more positive on moral values and ecology issues.

Conclusions

A review of the results of the practicum reveals that the anticipated student outcomes were substantially met. The writer found that using the Econet telecommunications network not only enabled the students to learn about this important technology but also focused the instruction in the resource room on important issues and topics. Writing skills focused on topics of world, national, and local concern, often dealing with ecology and environmental resources. The mechanics of writing were improved, especially for youngsters participating in the Kids' 93 project. They had to write letters of response to be sent to their "key pals" on the Kids' Cafe Conference. Reading skills connected learning across disciplines by using books, printouts, newspapers, and magazine materials on a variety of global issues. The material that was found on the network was as appropriate for the study of social studies (American History and Global Studies) as it was for science courses. Materials about human rights violations occurred as often as materials about toxic waste sites. Another area of deficit for many learning-disabled students is the need to develop organizational and higher-level thinking skills. Discussions about material found on the network, films, videos, and classroom journal topics fostered a sharing of information and insights among the students that enhanced their ability to analyze situations and look for solutions.

Learning-disabled students became more positive toward world issues, moral values, and ecological issues. Exposure to these issues through telecommunications and active teaching brought about a greater awareness of the possibilities for change in the future and thus a more positive attitude toward the issue itself. Students also reported that they had learned a lot about the environment and issues affecting the United States and the world.

Learning-disabled students also developed more positive attitudes about themselves and developed greater feelings of self-worth.

Dramatic Changes Occurred in Perceptions

The most dramatic changes occurred in the perceptions of students about ecology and the environment and about themselves. These were areas where direct and active teaching took place during this practicum project. This enhanced self-esteem and pride in themselves and in their resource room program brought about a willingness to speak about the project, write about it, and join in activities to promote it.

Student attitudes toward school at the end of the project were less positive for every group involved. Since each grade had such a small number of participants, it was felt that these data might not be representative of the whole group. Upon further analysis of the data, it was found that this trend seems to have been brought about by one or two students in each grade. The writer found that although teachers were individually interested in the telecommunications network as a concept, the reality is that few of them use computers or laser-disk technology in their classrooms. This is probably because they are not yet comfortable with technology in general. This practicum project found that as students were exposed to more concrete experiences, through taking field trips and becoming familiar with the work of environmental groups and scientists on the network, they became more knowledgeable and enthusiastic about their formal study of ecology and the environment.

References

Auchincloss, K., ed. 1992. "The Earth Summit: The Future Is Here." [Special Report]. *Newsweek* (June 1), 18-43.

Dodge, S. E., ed. 1992. "Disservice to the Parks." *National Parks* (December), 24-25.

Glenn, K. 1991. "Green Lies: An Environmental Issue." *Scholastic Scope* (September 13).

Hart, S. 1991. "UN Convention on the Rights of the Child." *School Psychology Review 20*:3 339–43.

Hoffman, P., ed. 1991. "The Celebration of Women in Science." [Special Issue]. *Discover 12* , 10.

Kadlecek, M. & B. Hogan. 1992. "Preventive medicine...for the environment." *The Conservationist* (November-December), 2-6.

Moser, D., ed. 1990. "Anniversary Issue: The Environment." *Smithsonian* (April).

Naar, J. 1990. *Design for a Livable Planet.* New York: Harper & Row.

Piel, J., ed. 1990. "Energy for Planet Earth." [Special Issue]. *Scientific American* (September).

Simons, M. 1993. "The European Community's Green Seals of Approval." *The New York Times* (April 11) E-5.

Steger, W. & J. Bowermaster. 1990. *Saving the Earth. A Citizen's Guide to Environmental Action.* New York: Alfred A. Knopf.

The EarthWorks Group. 1990. *50 Simple Things Kids Can Do to Save the Earth.* New York: Scholastic, Inc.

Weiner, J. 1990. *The Next One Hundred Years.* New York: Bantam Books.

Chapter 30

Building Ecological Thinking
in the Composition Classroom

Kelley M. Shull
Belleville Area College
Belleville, Illinois

Thinking and acting ecologically through education involves more than simply identifying and categorizing species and promoting slogans. Environmental consciousness generally relies on the elementary school child's recycling posters and on the environmental research done in chemistry, physics, geography, and biology laboratories on university campuses. While these all have their place, the environment cannot be restored or protected in the laboratory alone, as the effects of human impact on the environment are suffered by everyone and everything. Creating an environmentally healthy planet for human and nonhuman nature is a human affair that must involve a re-evaluation of our lifestyles and our impacts on ourselves and the nonhuman world. These are not necessarily scientific questions; they are social questions that require the evaluation of our belief systems and our communities and that must be explored through dialogue.

How Does Environmental Consciousness Grow?

The interaction between humans and the environment occurs in all aspects of everyday life, from pouring milk on cereal to brushing our teeth. To restrict environmental study to the sciences, therefore, is a false separation. Environmental consciousness must grow in humans, not through statistics and research, but rather through experiences that validate and strengthen the companionship between humans and the rest of nature (see McNamee, this volume). Anthony Weston defines ecological thinking as follows:

> the evolution of an interlinked system over time rather than in terms of separate and one-way interactions. Ideas, for example, not just ecosys-

tems, can be viewed in this way. Ethical ideas, in particular, are deeply interwoven with and dependent upon multiple contexts: other prevailing ideas and values, cultural institutions and practices, a vast range of experiences, and natural settings as well. (1965, 321)

If, as Weston claims, ideas can be viewed ecologically, then any and all academic subjects can be involved in environmental study. In fact, they must be involved if people are going to engage in ecological thinking that could change the current momentum toward the mass destruction of nature as we know it.

Generally, when we see environmental study in English departments, it consists of literature courses on nature writers or special-topic composition courses on environmental issues. Little has been done to look at composition ecologically and to explore what it means to write through an organic process that grows from the writer him/herself. I am still unsure of precisely what I mean by an "organic" writing process, but I come to this piece of writing knowing that my students fail to see their writing as extensions of their own lives and that they do not see writing as a natural process. From here, I am attempting to formulate an ecological approach to composition.

The parts of an essay are usually presented to students as complete and unchanging, and they are taught that simply learning the right places to plug in transitions and topic sentences makes a "good" paper. One of my students, named Rob, said, "We're like empty notebooks, and you're supposed to fill them up for us," as if he knew nothing about writing after his 50-some years. Rob's composition notebook is not empty; he is only convinced that it is, because the narratives he lives are not recognized or acceptable in the Academy. Although there are many theories of composition that refute this position, the discipline is generally still reduced to a "tell us what to write" task, where the most important aspects are the number of pages, run-on sentences, due dates, and double spacing. Not only do educators need to revolutionize their relationship with students and allow their voices to be heard in the classroom, but they also need to revolutionize the way composition is viewed as a subject.

Writing, Experience, and Ecological Thinking

The 1993 Scott Foresman, *Handbook for Writers* begins with a chapter on defining a "writing situation" and argues that writing must meet three goals:

"say something, to somebody, for some purpose." It continues by noting that "unless a piece of college writing meets these three goals, it doesn't really exist as writing. It's just an exercise" (12). What piece of writing doesn't meet those goals, and how is this division made? Even a grocery list or a note to my mother meets those goals. What writing isn't an exercise? But one important statement has been made, one which Rob has grasped quite well. The division might not be clear, but a division exists between "academic writing," which is "real" writing, and other writing, which is simply an "exercise." It is this division that leads Rob to believe he knows nothing about composition.

Viewing composition as a static, mechanistic process that separates us from experience not only increases anxiety and fear in writing students as they struggle to cross the dividing line into "academic writing" but it also ignores the connections and relationships that occur during the writing process. The *Handbook*, divided into sections resembling a drop menu on a computer, offers the message that if you simply learn the program enclosed in the "process menu," writing will become an easy, manageable task. But we write differently each time we sit down to compose as our audience, purpose, and needs change. Each person writes differently. So approaching writing as unchanging and mechanistic pretends that writing is independent of and external to our existence. Writing cannot be learned by purchasing *Writing for Dummies* at the local bookstore, like another Microsoft program, yet our current methods of teaching writing compartmentalize and kill the epistemic process of writing and violate the ecological aspects of the process.

Following Weston's claim noted previously, ecological thinking can be extended to the teaching of composition, as teachers/writers and students/writers begin to view writing as natural, following the forms, processes, and cycles that are linked to our physical environment, including our own bodies. Viewed as an ecosystem of experiences and ideas, writing is healthiest when diversity is able to flourish, growing through the interconnected process of individual cycles and feedback loops, while the writer creates a balanced system. Writing may then become an interaction among our emotions, experiences, situations, bodies, and language. Our relationship with language may then function as the primary foundation for writing, not the mastery of external rules, divisions, and forms. Of course, the writer must be nourished with guidance, good readers, and listeners, but not containment in an environment that is not conducive to growth.

Experiencing Writing as an Ecosystem

Ecosystems "consist of organisms, their environment, and all of the inter-actions that exist within. In other words, ecosystems are dynamic, interde-pendent, physical, and chemical" (Chiras 1994, 26). A wilderness pioneer community is the beginning of a sequential development in an ecosystem where species begin inhabiting new territory. Not necessarily because of damage, pioneer communities usually appear with weather changes or wa-ter-level changes that may, for example, expose previously hidden land to likely plants and animals. Lichens may find a cozy niche in newly exposed volcanic rock, and the outline for an ecosystem is in place. Species diversity is minimal at the outset, yet as the communities evolve, the species multi-ply, and a complex web of relationships evolves. Of course, there are no prescribed intentions outlined from beginning to end, with supporting de-tails mapped from the lichens' first growth, but simply a beginning to test out new habitat, grow seeds, and collect and make soil; the rest of the jour-ney forms as new growth forms.

Traditionally, composition instructors steer writers not into uncharted territories to develop intellectual ecosystems, not into new mental habitats for growth, but into restrictive prescribed territory that has been worn down, cut down, and killed as a resource for the Academy. Why is it so important for a "good writer" to create an outline first? Does this outline rope off territories that could create new ideas or new directions and produce a wider lens for experiencing the world? As composition instructors lead student writers down the same outlined trail, there is little space for growth, much like a hiking area that is surrounded by parking lots and trampled by hun-dreds of feet daily. It becomes worn out, dead, no longer encouraging growth. Hikes through these areas fail to arouse senses and emotions, and soon you no longer want or have a need to hike. My writing students, whether or not they have mastered it, have been taught writing for twelve or so years, and they have no excitement or interest anymore; writing is like a hike through a stressed-out, dead environment. Their senses are numb to the experience, and usually this lack of interest is blamed on them. They are stupid; they are lazy; they don't care—but a stressed-out system is not simply lazy, since options for growth are constantly trampled, and any attempt for growth is a kind of suicide mission.

As Annie Dillard (1989) suggests:

When you write, you lay out a line of words. You wield it, and it digs a path you follow soon to find yourself deep in new territory. Is it a dead end, or have you located a real subject? (3)

In composition, as in those overhiked wilderness areas, we need to begin fostering pioneer communities—nourishing new growth—so that students can develop diverse communities/writings and diverse forms of expression. Form becomes an integral component of a developing ecosystem, not a prescribed form, but form that takes shape with the growth of the sumac trees and raspberry bushes, the ideas and experiences. From an ecosystems perspective, form becomes something that grows as the community/writing forms. It is no longer something artificial. An organic form arises, "where content and form are in a state of dynamic interaction; the understanding of whether an experience is a linear sequence or a constellation that radiates out from and in to a central focus or axis, for instance, is discoverable only in the work, not before it" (Levertov 1973, 314). In teaching composition, instructors rarely allow form to take shape. Instead, they prescribe patterns, usually one (even though it may have slight variations). The *Handbook* calls it the "commitment and response pattern" (42), which we all know includes the introduction; arguments one, two, three, with examples and illustrations; and a conclusion that restates the introduction.

Experiencing Writing as Growth

My students have been taught this structure, have memorized it, and attempt to imitate it in their writing regularly. The result is usually disjointed essays with vocabulary, grammar, and sequencing problems connected vaguely with a broad generalization. Yet when I ask them to write a letter or a response that's not called an "assignment," their writing is typically clearer and more concise, with fewer grammatical difficulties, and the range of voices is tremendous. As a result, we spend many class sessions discussing forms in writing, and then they choose their own form. Usually, most of them continue fighting the ordained form, but some move beyond those restrictions. They actually begin hiking through new mental territories, experiencing writing as growth. We are inspired by a writer on writing:

You will know tomorrow, or this time next year. You make the path boldly and follow it fearfully. You go where the path leads….(Dillard 1989, 3)

The cycles of development from pioneer to intermediate communities in an ecosystem are constantly changing, readjusting, and re-creating themselves as they become mature and balanced. Within the mature system, elements—temperature, volcanic eruptions, hurricanes—constantly demand resistance, rebalancing, and recycling, while smaller communities find their own niches. A diverse community is not fixed but is continually beginning, and beginning, and beginning. A fallen tree is food for insects, for home for ground squirrels, and turns to soil for saplings. With each sentence I write, I am beginning a new chain, a chain of words, thoughts, and directions connected to others. I return to change the chains as new elements become important, as new demands surface, and as I live new experiences. Even as I become part of a mature community, simultaneously I am beginning new pioneer communities with other writers and within myself as I begin another paragraph, essay, or poem.

Ecofeminist Perspectives on Writing

In building a more ecological community of writers, our struggles rest in the dirt—in the struggles of being connected to ourselves and nature. As we learn to think and write ecologically—through examining connections, looking at relationships, and demystifying names and images of the human and the nonhuman—our actions must coincide. Nonhuman nature cannot stand together and revolt in the human sense. Therefore, by thinking and acting ecologically, we, "in theory and practice, attempt to speak for Nature—the 'other' that has no voice and is not conceived of subjectively in our civilization" (King 1989, 20). We must begin to recognize that many "voices" that are not human (but rather "whale" or "tree frog") speak through different senses within different contexts that are just as silenced as the voices of oppressed humans. Susan Griffin (1982) describes the unnamed voices:

> Behind naming, beneath words, is something else. An existence named, unnamed, and unnameable. We give the grass a name and earth a name. We say grass and earth are separate. We know this because we can pull the grass free from the earth and see its separate roots—but when the grass is free, it dies. (190)

It is difficult to speak of liberation for nonhumans using human terms, as we are restricted by human language to human experiences, emotions, and categories. It is important, I think, to recognize this limitation in our relationship with nonhuman nature, which creates for us the challenge of learning to create relationships that are not bound by human language. The voices of nonhuman nature, if we can use the human term "voices," communicate through our physical perceptions, our physical bodies utilizing our senses—the tastes, smells, sights, sounds, and touches—that humans tend to undervalue or ignore as forms of conversation. All of us have memories of a specific smell or taste that connects us with the past, but how can these experiences be expanded and connected to our relationships with nonhuman nature? My question is, how can we learn to connect, to have relationships, with nature that rely on more than verbal communication? I think we must begin by having this kind of relationship within ourselves, and writing can be this kind of conversation if it is approached as an organic process that grows through the self. Once we begin to have connected relationships with nature based on our physical perceptions, I believe we can begin to experience the world through connections rather than the delusions of which Griffin speaks. Foundations can be built through writing, and we can begin to experience writing in ways that exist in harmony with the natural world. We must learn to write in reflection about our actions and interactions with nature—human and nonhuman, to recognize the contradictions thrust upon us by the media, and to develop new ways of interacting with the world. Therefore, writing becomes a source of reflection, a process of reflection leading to, analyzing, and evaluating human actions and interactions.

Writers and teachers of writing must begin to teach students that writing is an ecological process, the development of a natural system of experiences and ideas, rather than a motionless, dead form that needs fixing and adjusting to become an "A." The lack of pioneer communities of writers and the weeding out of pioneer communities in English departments continue to break down the strength of composition as a valuable, imaginative art. Teachers of writing are becoming servants to business departments demanding dead prose. Fatal errors and punctuation have become obsessions, as we no longer have new growth to keep us enthralled. Aren't we sacrificing students when we teach them that errors are fatal?, that a comma is more vital than a new idea? If we continue to regard writing as motionless and static, fatal errors may be very fatal to the art of writing. We must continually traverse new territory and encourage students to do the same, so that readers and writers become part of an "interrelated web of language and an ecology of communication" and feel a greater connectedness in the interdependence of human and nonhuman nature (McAndrews 1991, 6). We must allow ecosystems to flourish within our own writing and in students' writing so that writing becomes a way of knowing, connecting and engaging in growth within ourselves and with our human and nonhuman environment(s).

References

Chiras, Daniel. 1994. *Environmental Science: Action for A Sustainable Future.* 4th ed. Redwood City: Benjamin/Cummings.

Dillard, Annie. 1989. *Teaching a Stone to Talk: Expeditions and Encounters.* New York: Harper & Row.

Griffin, Susan. 1982. "The Way of Ideology" In *Made From This Earth: American Women and Nature.* New York: Harper & Row.

Hairston, Maxine & John J. Ruskiewicz. 1993. *The Scott Foresman Handbook for Writers.* 3rd ed. New York: Harper.

King, Ynestra. 1989. "The Ecology of Feminism and the Feminism of Ecology." In Judith Plant, ed. *Healing the Wounds: The Promise of Ecofeminism.* Philadelphia: New Society Books, 18-28.

Levertov, Denise. 1973. "Some Notes on Organic Form." In Donald Merriam Allen, ed. *Poetics of the New American Poetry.* New York: Grove Press.

McAndrews, Donald. 1991. "Ecofeminism and the Teaching of Writing." Paper presented at the Conference on College Composition and Communication. Boston, MA. (March 21).

Weston, Anthony. 1992. "Before Environmental Ethics." *Environmental Ethics 14,* 321-33.

Chapter 31

The Greening of Academe:
A Provost's Perspective

Rosanne Wille
Lehman College
The City University of New York

As we approach the 21st century, higher education is being challenged to take more responsibility for preparing graduates to make an impact on their environment. According to John Vidal, reporting in the *Manchester Guardian,* three reports published early in 1997 point to "irresponsible and shortsighted governments" that are "pushing the world rapidly towards environmental and economic disaster by spending billions on the destruction of land, oceans, and the atmosphere" (1997). These reports—the Worldwatch Institute's "State of the World" (Brown, 1997), the "Panel on Sustainable Development" convened by Prime Minister John Major in the United Kingdom, and a report from Nairobi by Elizabeth Dowdeswell, Director of the U.N. Environmental Agency—warned diplomats from 100 countries that politicians were not taking seriously the eco-catastrophe being faced in Africa. Citing McKibben (1995), Kevles warns that, while it is obvious that environmentalism will have its effect on educational offerings, the impact is expected also to be on business (see Ruhl, this volume) as well as on the personal lifestyles of the next generation of active consumers (see Piorkowsky, this volume).

Colleges are being asked to prepare graduates with analytical and critical thinking skills, strong communication and technological skills while, at the same time, preparing them for active participation in a rapidly changing environment with a commitment to maintaining the integrity of our global ecosystem. These new demands present a daunting challenge as some members of the financially strapped academic community must shift resources away from some traditional areas (classics, for example) to focus on pragmatic curricula that provide realistic and relevant information for a society undergoing rapid social and technological change. Moreover, academics are being charged with the consciousness-raising of students toward the end result of sharing a moral and ethical commitment as responsible citizens of a democracy which includes maintaining environmental quality.

As Jeffrey C. Alexander has noted:

> the disciplinary basis of undergraduate education is not rational. It emerged for historical reasons that had nothing to do with pedagogy, and functions today primarily to support the creation, evaluation, and maintenance of new knowledge by scholars. But the university's second function, of course, is to help young people become more sophisticated thinkers, more responsible citizens, and more fully realized human beings. (B3)

Changing Emphases in Undergraduate Education

In the 1990s the undergraduate college degree is expected to carry with it the assurance of a secure niche in the workplace for the college graduate and the assurance of a well prepared employee for multiple workforces and changing workplaces. Laying the groundwork for lifelong learning and expected career changes are now major concerns of higher education. This causes faculty and administrators to re-examine and redefine the essential elements of a liberal education. With emerging prospects of a broad conceptualization of globalization and environmentalism comes the necessity for curriculum revision in the disciplines as well as in the basic general education requirements of colleges and universities. This comes at a time when colleges and universities are being urged to adopt a business model of accountability for faculty and administrators that identifies and treats students who, upon graduation, can be considered "products." Since budgets in higher education are enrollment driven, the fierce competition for students has encouraged a concomitant entrepreneurial spirit in faculty and academic administrators hitherto unknown and unprecedented. In this new climate, the traditional tensions between the liberal arts and the professional programs diminish in scope and importance as all disciplines are undertaking change based upon the current impact of globalization and environmentalism.

At the same time, a changing student population includes: (1) a more mature group coming to college with significant life experience in the United States, (2) increasing numbers of recent immigrants from all over the world, and (3) increasing numbers of native-born minority students pursuing higher education. Thus, the complexity and diversity of the undergraduate student body must be considered when making the needed changes because the student body itself is representative of a global perspective that serves as a medium for introducing the theme of environmentalism in every academic discipline.

Changes in the Canon

How does a changed academic climate alter our definition of the canon? What is needed to meet nontraditional expectations of a new body of students? Most important of all, what will become of colleges and universities if they move too slowly to make appropriate changes? How can a creative approach to curricula assure a positive outcome so that future generations will be afforded the security of a safe and healthy environment? And lastly, what is the most appropriate role of the Chief Academic Officer of a college in the change process?

Recognizing the rapidly changing nature of the student body, faculty have spent the last decade changing curricula to recognize multiculturalism and cultural diversity. Thus, we have, in effect, begun to change the canon.

The Role of the Provost in the American Academy

In the United States, the Provost serves a dual role in the higher education system. This role is one of balancing a decreasing budget with multiple competing needs of the academic and professional departments. As Dean of the Faculties, the Provost plays a pivotal role in curriculum change and development. But, in order to be effective, the role must be that of a facilitator, rather than a director of change. The Provost facilitates faculty interest and involvement and may suggest possible changes but he/she is not likely to be successful if a "Provost's Plan" is presented to the faculty as a *fait accompli* to be implemented without their input. Thus, the success of a college's efforts to respond to the call for environmentalism in the curriculum depends heavily on a Provost's relationship with academic deans, faculty, and other senior administrators. While stimulating change and anticipating changed curriculum imperatives, the Provost must be in a position to fund the planning activities necessary for collaborative work among the faculty as well as for the actual implementation of new curriculum offerings. Recognizing the lack of university positions available for doctorally prepared graduates in the traditional liberal arts, curricular changes are being made and priorities shifted with record breaking speed. Herein lies the tension between the "ideal" and the "real." This is as true for changes that reflect an environmental focus as for any other curriculum innovation.

Once a curriculum is approved, faculty participation becomes an issue. This also involves the recruitment and appointment of new faculty in appropriate departments, and (given the "territoriality" of academic departments) touches on the sensitive issue of joint appointments. It is important to work from a systems perspective and to have a collegewide effort involved in the change process. If college-wide cooperation is not successful, the Provost faces the difficulty of making choices among multiple new programs and positions on the part of many competing academic departments and divisions. This inevitably results in some departments and divisions being "left out" and feeling disenfranchised after having invested considerable time and effort. Thus, the key to success is to foster mutually beneficial collaborative relationships among academic disciplines. This then becomes the culture of the college *prior to* attempting overall incorporation of environmental studies into the college curriculum as a whole.

Curriculum in a Time of Imperatives for the 21st Century

Funding for faculty efforts in curriculum change of any kind is complicated by the fact that—worldwide—a great many colleges and universities are in financial difficulty. Therefore, funding choices are most often made by assessing the number of students and faculty who can benefit from proposed changes. At Lehman College we have created a culture of collaboration and cooperation as a means of survival in the complex City University of New York system. Three years ago, the RELATE Committee began its work to foster a closer working relationship and an atmosphere of mutual respect between the teacher education departments and the liberal arts, natural, and social sciences. The success of this committee is apparent in the new found working relationship among faculty from the different disciplines. For example, a proposed Urban Environmental Sciences and Policy Program, involving departments in the natural sciences, social sciences and humanities, is gaining support from faculty, students, and administrators.

While CUNY budget deficits have forced changes in the working relationships among faculty and administrators at Lehman, they have also caused faculty from different disciplines to work together in a collaborative-cooperative mode that has effectively eliminated the territoriality that resisted such cooperation in the past. In effect, the rules of the game have changed such that the reward system, while smaller, is directed to joint curricular efforts rather than to single departments. In other words, the Academy itself became more "ecological" in character. This came about through the active participation of senior faculty; it has now become the norm for the college. In doing this out of necessity and from the identification of shared interests, a new academic culture has managed to reenergize and reawaken the creative strengths of faculty in most of the 25 academic departments at Lehman.

As we continue to work toward the "greening" of the curriculum, we are being forced to question (and perhaps to redefine) the traditional labels of liberal arts, natural sciences, social sciences, education, and health. The barriers created by such labels limit thinking in such a way that environmental and global issues, which span all disciplines, create "border crossings" that serve to blur traditional disciplinary boundaries. To be successful, environmental education must take these complex relationships into account.

The time has come to incorporate EE across the college curriculum. This is occurring in all disciplines at many levels and, as the contributors to this volume attest, in many unexpected areas including the arts. We have responded to the changing needs of society and our students. We have also begun to respond to the new demands of the workforce by the incorporation of technology, communication and critical thinking skills into the curriculum. We have begun to rise to the challenge of consumer demand and to shed our collective need to clone ourselves through the classroom and laboratory contact with the next generation of potential scholars.

Competition for Scarce Resources

About two million scholarly articles in science and engineering alone are published every year by 72,000 journals (Denning, 27). College libraries can afford to subscribe to only a small fraction of these journals. Consequently, the vast majority of articles go unread or read by very few, and most are never cited by another author. It can be concluded, therefore, that certain aspects of the tenure and promotion criteria themselves have led to a weakened university system. Even the National Science Foundation is beginning to focus, not on research itself, but on the relationship of research to curricula and to classroom teaching-learning processes. These changes are taking place in a system notorious for its rigidity and resistance to change. Joy Palmer (this volume) makes the distinction between teaching *about* the environment by transmission of knowledge; teaching *in* the environment through fieldwork; and teaching *for* the environment with promotion of the caretaker ethic.

Like writing, critical thinking and technology, environmental and global issues are equally appropriate in the natural sciences, social sciences and
the arts and humanities. We cannot forget that curricular innovation—however desirable—involves economic cost, political trade-offs, psychological
consequences, and social effects. It is clearly understood that environmental issues are social, political and economic in nature. Thus, environmental
topics fit naturally across the curriculum with more ease than even
multiculturalism.

Many instructional materials have been designed to provide information and to focus solely on fostering environmental awareness rather than
participation. Consequently, few individuals are exposed to or expected to
acquire skills necessary for environmental stewardship and empowerment
(see McConney, et al. this volume). Even Rachael Carson's three periods of
writing demonstrate this natural progression: the nature-study era, the explicit concern, and the questions and values, problem solving and action
taking dimension (see Corcoran, this volume).

The fundamentals of EE belong in the primary and secondary grades
with the continuum of building in colleges and universities where the more
complex issues of political action and community problem-solving are addressed. Teacher education becomes critical in this context (see Oulton &
Scott and Semeraro, this volume).

Learning about the environment should begin well before primary
school. A basic ethic handed down from parent to child is the ideal that we
have yet to achieve. An ethic of care is the foundation of a caring ecological
relationship (see McNamee, this volume). Gender is not generally a strong
factor to be considered in environmental education. However, even today
when gender roles are likely to be blurred, women are most likely to assume
responsibility for incorporating ecological standards and environmental
respect into the usual household routines (see Methfessel, this volume).
Women generally have the responsibility to instill ecological values in the
pre-school-aged child. However, this early education can only take place if
the mother has the interest and opportunity necessary to achieve the appropriate knowledge base.

Thus, primary and secondary schools need to make special efforts to
attract girls to science classes as learners and participants. Learning about
the environment, particularly in the primary grades, is predominantly a
matter of providing information through basic science classes that focus on
the wonder of all living things. If interest is sparked in the pre-school envi-

ronment, the stage is set for this active learning in primary and secondary schools. This makes environmental education in teacher education mandatory.

Conclusion

Since college curricula must address the multiple complexities of general education and disciplinary requirements, it stands to reason that learning *about* and learning *in* the environment must first take place at the primary and secondary educational levels (see Mortari, this volume). This leaves learning to take responsibility for the environment to higher education. Of course, this educational progression calls for collaborative planning among colleges and primary and secondary schools. Environmental education poses a complex set of issues since at the higher levels it becomes, of necessity, an interdisciplinary responsibility. As environmental education takes its proper place in the curriculum, high schools must prepare graduates with the ability to engage in informed debate about environmental issues with a sound, accurate subject knowledge base. This knowledge base will allow for the informed exploration of the scientific, social, and economic issues that will lead to internalized and informed environmental ethical value systems for college graduates. This is particularly important in the area of teacher education in order to provide the knowledgeable professionals necessary to provide the foundation for college level work and so to make the cycle complete.

References

Alexander, Jeffrey C. 1993. "The Irrational Disciplinarity of undergraduate Education." *The Chronicle of Higher Education* (December 1), B3.

Brown, Lester. 1994. *The State of the World.*

Denning, Peter. 1996. "Business Designs for the New University. *Educom Review* (November/December), 21-30.

Kevles, Daniel J. 1997. "Endangered Environmentalists." [Review essay.] *The New York Review of Books* (February 20), 30-33.

McKibben, Bill. 1995. *Hope, Human and Wild.* New York: Little Brown.

Vidal, John. 1997. "World Turning Blind Eye to Catastrophe. *The Economist* (Feb. 2), 1

Chapter 32

"Media-ting" the Environment: A Case Study of the College/Community Relationship

Anne D. Perryman
Lehman College
The City University of New York

N o one today underestimates the importance of effective media relations for an academic institution. This chapter presents a case
study of how Lehman College communicated the vital relationship between a campus and its community and of the college's role as a
cultural resource in an urban ecology.

In the interest of public image, clients are often advised to "go with
what you've got." When I first saw Lehman College in the spring of 1988, I
was unprepared for the extraordinary beauty of the campus—its handsome
buildings, grassy knolls, old sycamore trees, and the stunning reflection of
Gothic towers across the sleek white surface of the Concert Hall in late af-
ternoon sunlight. Despite the close proximity of an elevated train, there is a
sense of peacefulness and a natural harmony about the place. As the new
director of media relations and publications, I saw the campus as a treasure
that we should talk more about and show off in photographs in all of our
publications and public relations initiatives. When *The New York Times* described Lehman as "the most attractive of the CUNY colleges…with some
of its finest facilities," it was as if the personal opinion of many of us on
campus had received validation from the nation's foremost paper of record.
A gorgeous campus—tucked away in the northwest corner of the Bronx—
was what we had, and we needed to make the most of that.[1]

But we had a problem. In 1988, it was impossible to reach Lehman's
campus either by car or by public transportation without passing through a
degraded area that was city-owned. This sad state of affairs had come upon
the neighborhood gradually, and most people seemed to accept the erosion
of municipal services. After all, during the previous two decades, New York

(like many of the nation's great cities) had been hit by recessions, an eroding infrastructure, rapidly shifting demographics, and rising unemployment, especially among unskilled youth in the inner city.

As the health, housing, education, and workplace needs of new generations of New Yorkers exploded, federal and state aid for the City were reduced and sometimes eliminated. And when New York City suffers, the pain felt in the Bronx—where so many residents are struggling financially—is disproportionate. In light of these daunting circumstances, littered streets did not seem to be anyone's priority.

Ring Around the Campus: An Environmental Problem

Clearly, the unsightly ring around Lehman's campus was part and parcel of deeper problems in New York City. To the west, for example, public green spaces around the Jerome Park Reservoir had become a dumping ground after the city's downsized Department of Environmental Protection (DEP) built a tall fence around the water and, in effect, turned inward—out of reach and out of touch with its neighbors. Clearly, it was the DEP's job, first and foremost, to protect the city's water supply. In the process, a beautiful body of water, once an important community amenity, almost disappeared behind a perimeter of thick weeds and trash.

To the east, at the New York Transit Authority's Number 4 line train station at Bedford Park Boulevard, the cleanup services of two city departments failed to connect. Transit Authority porters cleaned inside of the station and along the elevated tracks; Sanitation Department workers swept the street. Meanwhile, riders tossed their trash over the sides of a veranda onto a growing mound of litter; a no-man's land of debris was created directly in front of the heavily used station.

To make matters worse, the Transit Authority, as part of its campaign to remove graffiti from the subway, had installed rolling razor wire on top of the fences surrounding its rail yards under the Bedford Park Boulevard Bridge. The bridge is the main approach to Lehman College and the adjacent Bronx High School of Science from two train stations—the Number 4 elevated line and the "D" train, a subway line two blocks away. It is no exaggeration to say that the installation of the razor wire made Lehman's campus look like a maximum security prison, from the outside at least.

In the spring of 1991, the entire community was thrown for another loss. The Bedford Park Boulevard Bridge was one of two New York City bridges to be condemned and closed because the city could not make costly repairs. Huge concrete barricades were installed at both ends of the bridge. They were immediately hit with graffiti. Sanitation trucks could no longer get onto the bridge to sweep it; the Department of Transportation told us it lacked the resources to provide regular maintenance. Thus, the 60-year-old bridge became a wind tunnel, with litter collecting in the corners along the barricades and large puddles forming after a hard rain. Although it was closed to vehicles, the bridge continued to be used daily by thousands of pedestrians, including students at Lehman College and the Bronx High School of Science as well as residents of the community. What could be done to improve these unsightly public spaces? What is a city college's responsibility to resist blight and decay? Is there a college/community/environmental compact? We think there is.

Making a Difference

Lehman College did not develop a master plan to improve the troubled environment around its campus and restore the ecology of the community, although we can almost make it sound as if we had. Our efforts since 1988, when the college became the first Transit Authority Adopt-a-Station sponsor in the Bronx, have been remarkably successful (see Fernández, this volume). The truth is that we all had other jobs to do (the jobs for which we were being paid); our volunteer community-environment activities often had to wait on the back burner. Though we tried to make progress on chronic issues, typically we were forced to deal with a string of small problems as they arose (from an abandoned car in the streets to scary dogs running unleashed on the bridge). We never enjoyed the luxury of simply moving forward; it was always two steps forward, one step back—and sometimes two or three steps back! Lehman's neighborhood environmental efforts have been a continual struggle, a media relations campaign, and a volunteer movement at the same time. Sometimes volunteers worked alone while passersby commented: "It's hopeless," "You're swimming upstream," "Give up—you can't win." But more often than not, community residents cheered on the volunteers and pitched in themselves to help.

In 1990, our fledgling campaign gained a powerful ally: the new President of Lehman College, Ricardo R. Fernández. Fresh from the green campuses of the University of Wisconsin, he was appalled by the lack of city services; he agreed that it was disrespectful to our students to be forced to walk through trash to attend college classes. He encouraged college volunteers to continue their efforts (which by then also included campus-wide white paper recyling), and he began his own campaign. He and the presidents of several major Bronx institutions—Fordham University, the New York Botanical Garden, the Wildlife Conservation Society (Bronx Zoo), Bronx Community College, and the Montefiore Medical Center—met with the Chairman of the Metropolitan Transit Authority to discuss public transportation issues and problems. Because of our Adopt-a-Station involvement, President Fernández was able to take to that meeting a list of specific, doable requests, which the MTA chairman passed along and which resulted in immediate action.

Between 1988 and the present, Lehman College volunteers, in collaboration with other schools, community groups, municipal agencies, and environmental organizations, have implemented more than a dozen environmental projects around the campus (see page 349). The college and the City Volunteer Corps have worked together on seven different projects. In recognition of these efforts, the college received "Molly Parnis Neighborhood Beautification Awards" and grants from the Citizens Committee for New York City in 1988, 1989, 1990, and 1992. Lehman's Office of Media Relations and Publications documented and publicized each project with photographs and press releases to local papers. Over the years, Lehman students and employees—including the president of the college—have become a familiar sight picking up litter, raking leaves, planting flowers, making a difference in the community.

Environmental Issues for the 21st Century: Organizing a Conference

How could this commitment to environmental activism be extended to Lehman's academic community? The college had staged a campus-wide effort for the "Mayakovsky Centennial" in 1993. Why not use the campus as a setting for a conference on the environment? This plan posed challenges and presented opportunities to the entire college community.

Reaching out to environmental educators and activists followed a natural direction when President Fernández designated 1994-95 as "Earth Year at Lehman College." Organizing a week-long conference on "Environmental Issues for the 21st Century" gave the college an opportunity to bring together a unique group of scholars who viewed environmental issues from diverse disciplinary perspectives. Chaired by Professor Patricia J. Thompson and supported by the college's Provost and a dedicated committee (but with no budget earmarked for the event), the effort began with a "call for papers" in the *Chronicle of Higher Education.*

The Keynote Connection

Again, the motto "go with what you've got" came to mind as we began to plan the conference: Who in our university system could be called on as keynoters? Certainly, we needed people who had name recognition and the respect of environmentalists. As it happens, Professor Barry Commoner, whom many consider to be the "father of the environmental movement," is director of the Center for the Biology of Natural Systems at Queens College, a sister institution in the CUNY system. The eminent author, biologist, and environmental crusader graciously agreed to give the conference keynote on "What *Is* the Environmental Imperative?" For the international visitors to the conference, Dr. Commoner's opening remarks were especially significant, because they were all familiar with his groundbreaking work, and this was their first opportunity to meet him personally. A professor at yet another CUNY institution, Michio Kaku of City College, is a leading theoretical nuclear physicist, radio talk show host, and an outspoken critic of U.S. energy policy. Professor Kaku agreed to give a plenary session keynote; it was a riveting address on "The Collapsing Environment: Is There a Point of No Return?"

Four public figures well-known to environmentalists also accepted our invitation to speak. Robert F. Kennedy, Jr., whose reputation as a defender of the environment stems from a series of successful legal actions, spoke to the closing plenary on "Environmental Meltdown: Is This Our Contract with the Future?" As chief prosecuting attorney for the Hudson Riverkeeper, senior attorney for the Natural Resources Defense Council, and supervising attorney for the Environmental Litigation Clinic at Pace University School of Law, Kennedy has led the fight to protect New York City's water supply. Indeed, the watershed agreement he negotiated on behalf of environmentalists and New York City is regarded as an international model in stakeholder-consensus negotiations and sustainable development. Because our campus is located next to a reservoir where a huge water filtration plant has been proposed, conference participants and community residents alike were interested in Mr. Kennedy's impassioned comments. We were delighted also by the willingness of another extremely busy individual and powerful speaker, The Hon. Bella Abzug, to speak to the conference on the question of "Environmental Impacts on Women's Health." The former New York Congresswoman now heads the activist group WEDO (Women's Environmental and Development Organization). The Hon. Orin Lehman, a prominent New Yorker, friend of the college, and for many years Commissioner of the New York State Office of Parks, Recreation, and Historic Preservation, spoke on "Public Issues in Parks and Historical Preservation." To keynote the international perspective, Patricia Mische, author and researcher on global environmental issues and director of Global Education Associates, agreed to speak on "Ecological Security in an Interdependent World." These prominent speakers waived their normal fees to address Lehman's conference without honoraria.

Panelists Discuss the Military Impact on the Environment

Rarely heard at an academic forum is the military perspective. The college invited a panel to address "The Impact of Military Activities on the Environment." This was chaired by Victor Sidel, M.D., of the Montefiore Medical Center, Albert Einstein College of Medicine, a neighboring institution in the northwest Bronx. The panel also included Michael Renner, senior researcher from the Worldwatch Institute; Capt. Jack Ahart, a professor at the Naval War College (see Ahart, this volume), and Professor Patricio Lerzundi of Lehman College, representing the International Association of University Presidents/UN Commission on Disarmament and Peace Education. Dr. L. Eudora Pettigrew, president of the State University of New York College at Old Westbury and co-chair of the IAUP/UN Joint Committee on Arms Control Education, joined the panel as a discussant.

Communicating the Environmental Imperative: A Distinguished Panel

In response to the call for papers, the Committee received a proposal from David B. Sachsman, who holds the George R. West Chair of Excellence in Communication and Public Affairs at the University of Tennessee-Chattanooga. Dr. Sachsman made us aware of the newly emerging role of environmental journalism in improving the public's understanding of environmental issues (see Sachsman, this volume). This fortuitous contact led to an invited symposium with the provocative title "Environmental Risk Reporting: Do the Media Give a Damn About the Environment?" which Sachsman moderated. With generous travel support from The Overbrook

Foundation, TWA, and Delta Airlines, the college was able to invite four of the nation's leading environmental journalists to participate in a morning symposium at Lehman College and in an additional evening symposium at the Columbia University Graduate School of Journalism, thus expanding the college's community outreach from the Bronx to Manhattan. The journalists:

• Philip Shabecoff was *The New York Times'* award-winning principal environmental correspondent from 1977-91, and is publisher of the environmental news service *Green Wire*. He is the author of *A Fierce Green Fire: The American Environmental Movement* and *A New Name for Peace: Environment, Development and Democracy*. (Shabecoff is also an alumnus of our institution).

• Jim Detjen, a 20-year veteran of environmental reporting, has received more than 40 awards for his investigative reports ranging from PCBs in the Hudson River to the Chernobyl nuclear accident for the *Philadelphia Inquirer, Louisville Courier-Journal,* and the *Poughkeepsie Journal*. Detjen now holds the prestigious Knight Chair of Environmental Journalism at Michigan State University.

• Rae Tyson, an environmental writer/editor for *USA Today* and the author of *Kidsafe*, has won major awards for his Love Canal reporting for the *Niagara Falls Gazette*.

• Emilia Askari, an environmental writer for the *Detroit Free Press*, is noted for her reporting on pollution in the Great Lakes and on the environmental justice movement. Askari has travelled on assignment from the rainforests of the Amazon to the glaciers of Tierra del Fuego. She was, at that time, president of the Society of Environmental Journalists.

• Teya Ryan, executive producer of "TalkBack Live" for CNN and co-executive producer of CNN's Environmental Unit, had received an Emmy Award for the CNN production "In Nature's Wake."

The audiences for these symposia included large numbers of students (many of them reporters on high school and college student newspapers) interested in the possibility of careers in environmental journalism. The symposium is an example of how a good suggestion from an unexpected source can enliven a conference. Assembling a panel such as this is an option for any conference planning committee.

Speakers on Current Issues

One of President Fernández's goals from the outset of the planning process was for this conference to deal with local issues in the Bronx community as well as national and international educational concerns. To give the keynote at a plenary session "In Our Own Backyard: Whose Job is It?" the Committee called on Michael Zamm, director of education for the New York City Council on the Environment and one of the founders of a public high school for environmental studies in New York City. He spoke on "Mobilizing the Community: A Grassroots Approach to Environmental Projects." Peggy Shepard, well known to local environmentalists as executive director of the West Harlem Environmental Action Coalition, spoke on "Organizing the Community for Environmental Equity." A presentation by Pierre Erville, of the Washington, DC Alliance to End Childhood Lead Poisoning, addressed "The Global Dimensions of Lead Poisoning." This was an especially relevant topic for the Bronx community, where many deteriorating buildings pose a lead poisoning threat to the health of children.

A Town Meeting on Noise

Lehman College professor emerita Arline Bronzaft, an environmental psychologist who has written widely about the impact of noise on learning (see Bronzaft, this volume), proposed to the planning Committee that the college sponsor a "Town Meeting on Noise" during the conference. The event was held at St. Philip Neri Parish Center a few blocks from the campus. The speakers included Dr. Bronzaft; John Dallas, founder and president of the Bronx Campaign for Peace and Quiet; and Nancy Nadler, director of the League for the Hard of Hearing's Noise Center. The speakers made concrete points concerning the impact of noise on the quality of life in urban neighborhoods. The "Town Meeting on Noise" attracted a large crowd of concerned citizens and led to more town meetings at which elected officials and members of the law enforcement community were called to account on noise-related issues.

Cultural Events

During the conference, the campus was alive with cultural events that brought the audience into contact with artistic presentations in various media special events. The special events included the following:

• A Chinese calligraphy demonstration, "Nature with the Stroke of a Pen," by Huang Tiechi, a visiting professor from the University of Shanghai, was held in the Lehman College Art Gallery.

• The Art Gallery also exhibited "The Art of Rigoberto Torres," life-size figurative sculptures that are a celebration of daily life on the streets of the South Bronx. Torres generously agreed to give a public demonstration of his artistic casting technique and to create a sculpture of one of the conference participants (a member of the City Volunteer Corps was the lucky winner of the sculpture lottery).

• The Lehman College Library, under its director, Professor Daniel Rubey, featured "Human Relationships with Space and Time," a photographic exhibition by environmental artists Terri Warpinski, Emily Gertz, Jerry Dell, and David Taylor that explored landscapes past, present, and future.

• Other events in the Library included a Children's Film Festival, organized by Bullfrog Films, a Children's Book Exhibition, and a special display of current environmental books.

• Two poetry readings were held in the Lehman College Art Gallery during the conference. A Poets Circle, sponsored by the Bronx Council for Environmental Quality, featured the SAF Forum Poets of Schyulkill County, Pennsylvania.

• Nearly 500 members of Bronx High School choruses—on campus for a choral festival—took time out to sing for conference participants in the campus oval.

The University's First Electronic Conference

Recognizing the national and international interest in environmental studies, Professor Sheila Smith-Hobson, with the assistance of Emily Gertz and EcoNet, mounted the City University's first E-Conference. The E-Conference invited Internet users to read and comment on ten conference papers in a unique contemporary context. The innovation was so successful, it ran for more than a month—and gave the college's "techno-academics" an opportunity to experiment with what will surely become an important feature of future scholarly meetings.

The Education Expo

As part of the conference, the Division of Education hosted an "Education Expo," organized by Mikki Weiss and Judith Guskin with support from Kimo

Kimokeo of the U.S. Forest Service. The halls were filled with colorful posters. Exhibits created by students and teachers in Bronx schools offered a mini "science fair." Workshops and curriculum showcases were organized for the school district's teachers and supervisors. The collection of posters by students at Bronx High School of Science was so impressive, it led to another exhibition curated by the U.S. Forest Service.

Conclusion

The Earth Year conference events I have described were not all planned in advance. They developed organically, much as Kelley Shull (this volume) speaks of creative writing. It was, in effect, an ecological process that began with a small group of enthusiastic people brainstorming ideas. From these ideas, we made connections, developed synergies, and created a positive energy that sustained us and enlivened the campus during the entire conference week. This case study is not intended to be a recipe for others to follow. But every college can adopt the motto "go with what you've got" to strengthen ties among academic and civic constituencies in their communities. Every college and every community has its share of scholars, artists, activists, journalists, teachers, and students from which to draw its own celebratory occasions. Bringing them together in common cause is also an environmental imperative for the 21st century.

Notes

1. This chapter is a revision of a paper, "Upgrading the Urban Environment: The Role of a Public College in a City University System," first presented at a roundtable at the annual meeting of the American Educational Research Association in New Orleans on April 8, 1994.

A College Campus and Its Community

Lehman College's contribution to environmental quality included some of these activities, many of which could be adapted to local conditions and available resources:

• *Neighborhood clean-ups in the spring and summer.* As many as 150 members of an honor society from a nearby high school (the Bronx High School of Science) participated, using clean-up tools and supplies from the Council on the Environment of New York City and We Care About New York, a community-based program affiliated with the Sanitation Department.

• *Spring planting of flowers and fall planting of bulbs.* Plant materials were contributed by the Parks Council and the Bronx Green-up Program of the New York Botanical Garden.

• *City Volunteer Corps projects.* Volunteers landscaped the front of the Bedford Park Boulevard train station, built benches, brought in topsoil, and planted a garden. The CVC is a New York City youth training program with an educational component.

• *Bedford Park Boulevard Bridge Walkway projects.* This effort received broad community support. The Bronx Borough President contributed 18 tree tubs, which allowed the Department of Transportation to remove its unsightly barricades. The City Volunteer Corps planted shrubs from the Bronx Green-up program in the tree tubs, and workers from the Bedford Park Car Wash are watering the shrubs.

• *The Jerome Park Conservancy.* This community-based greens advocacy group was formed in opposition to a Department of Environmental Protection proposal to build a water filtration plant at the Jerome Park Reservoir immediately adjacent to the Lehman campus. Lehman College President Ricardo R. Fernández became chairman of the Conservancy.

• *The President's Green Team.* This project involved students in campus recycling and neighborhood clean-up initiatives. The college gained credibility as an intrinsic part of the community. We not only "talked the talk;" environmentally speaking, we "walked the walk."

The Contributors

Jack Ahart is a retired Captain in the U.S. Navy and former Professor at the U.S. Naval War College in Newport, Rhode Island.

Dieter L. Böhn is Professor of Geography at the University of Würzburg in Germany. He serves on the "Environmental Station" board in Würzburg and is a recipient of the Bavarian Ministry of Environment's Medal of Honor.

Edward Boyes is Senior Lecturer in the Department of Education at the University of Liverpool, UK, where he is a member of the Environmental Education Research Unit.

Arline L. Bronzaft is Professor Emerita of Psychology at Lehman College, and an expert witness and consultant on noise. She has written chapters in the *Encyclopedia of Environmental Science, Engineering,* and *Noise and Health.* Her most recent book, *Top of the Class,* includes her research on noise and children's learning.

Patricia Carlson teaches chemistry and environmental science at New Trier High School in Winnetka, Illinois. Her publications include "Environmental Investigations: The Science Behind the Headlines" (*Science Teacher*), in which she presented a case study of an open-ended inquiry in technology.

Peter Blaze Corcoran is Associate Professor and Chair of the Department of Education at Bates College in Lewiston, Maine. His articles have appeared in *Environmental Education Research, Environmental Education,* and *Journal of Environmental Education.* He serves in 1997 as President of the North American Association for Environmental Education. He has been a visiting professor at two Australian universities and is on the editorial board of *International Research in Geographical and Environmental Education.* His scholarly interests include the significant life experiences leading to environmental education and spirituality in education.

Margery Cornwell is Associate Professor of English in the Department of English, Speech, and World Literature at the College of Staten Island, The City University of New York. Her long-time interest has been in environment and literature. She was editor of *Three Grey Geese: Modern Writing in Honor of Animals.*

Ricardo R. Fernández is President of Lehman College, The City University of New York, and Professor in the Department of Languages and Literatures. He is the Chair of the Jerome Park Conservancy, serves on the Wildlife Conservation Society Advisory Board on the Congo Rain Forest, and is a trustee of Wave Hill, a public garden and cultural center. Dr. Fernández is interested in public policy issues such as the role of public higher education in a multicultural society and the environment.

Jacqueline Gravanis is Professor of Home Economics and Education at Harokopio University in Athens. She was educated in the United States and Greece. She has published on food science and adolescent nutrition and has presented papers at conferences in the U.S., Israel, and Greece.

Philip B. Horton is Professor of Education at the University of Jos in Jos, Nigeria.

Tiechi Huang is Associate Professor in the Department of Chinese Language and Literature at Shanghai Teachers University in China. His research interests include comparative literature and contemporary literature and art of Western countries. He is an accomplished artist in Chinese painting, calligraphy, and seal carving.

Esther Zager Levine is a resource room teacher of the learning disabled at Valley Stream North High School in Franklin Square, New York. She has a special interest in the use of computers in educating special populations.

Amanda Woods McConney is Lecturer in Environmental Studies, College of Arts and Sciences, and Office for Continuing Professional Development, College of Education, Western Oregon State College in Monmouth. She is the author of the *Environmental Science Instructor's Manual* for Cunningham and Saigo's *Environmental Science: A Global Concern*, 4th ed. She was a 1991-92 Fellow of the Pew Charitable Trusts Program Initiative: Integrated Approaches to Training in Conservation and Sustainable Development.

Andrew A. McConney is Associate Research Professor in the Teaching Research Division at Western Oregon State College at Monmouth. He is a member of the Ecological Special Interest Group of the American Educational Research Association. He and Amanda Woods McConney have collaborated on a number of research projects in environmental education and presented papers at international consortia.

Abigail S. McNamee is Associate Professor of Child Development, Coordinator of the Graduate Program in Early Childhood, and Chair of the Department of Early Childhood and Elementary Education at Lehman College. She works as a child therapist and is currently completing her second doctorate in developmental psychology.

Gene McQuillan is Assistant Professor of English at Kingsborough Community College, The City University of New York, where he designed a new course, "The Literature of Adventure and Exploration." He was the convener of a conference "From the Frontier to la Frontera: American Women/American Landscapes." He is presently at work on a book manuscript, *Demanding a Different World: Essays on Travel, Tourism, and Adventure*.

Barbara Methfessel is Dr./Professor at the Pädagogische Hochschule in Heidelberg, Germany, where writes extensively on changing family roles and family resource management from a feminist perspective. She was a contributor to the volume *Hausarbeit und Bildung: Zur Didaktik der Haushaltslehre*.

Luigina Mortari is a Laureate in Pedagogy and Doctor of Pedagogical Research and Education Sciences on the faculty of the Institute of Educational Sciences at the University of Verona, where she is responsible for ecological education. She is the author of *Abitare con Sagezza la Terra: Forme Costitutive dell'educazione ecologica*.

C.R. Oulton is Senior Lecturer in the Department of Education at the University of Bath, UK, and deputy director of the Centre for Research in Environmental Education, Theory, and Practice. He serves as editor of the journal *Environmental Education Research*.

Joy A. Palmer is Deputy Dean of the Faculty of Social Sciences at the University of Durham, UK. She has published books and articles on environmental education and is currently working on a book of biographies of leaders in the movement for environmental education.

Anne D. Perryman, a photojournalist and former foreign correspondent, is Director of Media Relations and Publications at Lehman College. In 1992, she received the New York City Mayor's Award for Volunteer Service for her efforts to restore and beautify the Bedford Park Boulevard subway station in the Bronx.

Michael-Burkhard Piorkowsky is Dr./Professor at the Rheinische Friedrich-Wilhelms Universität where he teaches Consumer and Household Economics. He is Past President of the German Home Economics Association.

Kate Potter is a published poet who also led a successful resistance to a proposal to import and incinerate contaminated soil in her rural Pennsylvania community.

Frank E. Puin is an environmental activist and teacher. He is Director of the Center for Social and Ecological Technological Research in Hannover, Germany. He has organized student cooperatives and volunteer activities at Fachhochschule Fulda, in Fulda.

Gaoyin Qian is Assistant Professor of Reading in the Division of Education at Lehman College. His research interests include conceptual learning in secondary school students' learning from science text and literacy development among young children from immigrant families.

Glenn D. Ruhl is Education Program Manager for business-sponsored programs of the innovative Environmental Services Association of Alberta, Canada.

David B. Sachsman holds the George R. West, Jr. Chair of Excellence in Communication and Public Affairs at the University of Tennessee at Chattanooga at the rank of Professor. He is known for his research and scholarly activities in environmental communication and environmental risk reporting and for the three editions of *Media: An Introductory Analysis of American Mass Communications*, which he worte with Peter M. Sandman and David M. Rubin.

Hiltraud Schmidt-Waldherr is Dr./Professor at the Fachhochschule Fulda in Fulda, Germany. Her research emphasis is in household economics and human ecology. She has written and lectured widely on feminism and household economics, gender theory, and human ecology. She was co-editor of *FrauenSozialKunde: Wandel und Differenzierung von Lebensformen und Bewusstsein.*

William A.H. Scott is Senior Lecturer in the Department of Education at The University of Bath, UK, where he has served as Director of the Center for Research in Environmental Education and Chair of the Editorial Board of *Environmental Education Reseach.*

Raffaella Semeraro is Professor in the Department of Educational Sciences at the University of Padua, Italy, where she teaches courses in theories and methods of instructional design and assessment.

David W. Shapiro is Assistant Professor of Corporate Communication in the Roy H. Park School of Communications at Ithaca College in Ithaca, New York. He was co-editor of a special issue of *Educational Technology:* "Perspectives on Change" (January-February, 1996). He participated in a teacher workshop on environmental ethics at the Four Corners School of Outdoor Education in Monticello, Utah (summer 1996). He is co-chair of the Board of Directors for CRESP (Center for Religion, Ethics and Social Policy), an action-based educational organization that is affiliated with Cornell University.

Kelley M. Shull is a writing instructor at Belleville Area College in Illinois and an organizer for the National Wildlife Federation's Campus Ecology Program. She has worked with the Student Environmental Action Coalition (SEAC) and other campus organizations confronting women's issues and environmental justice.

N.J. Smith-Sebasto is Assistant Professor and State Extension Specialist at the University of Illinois at Urbana-Champaign, where he teaches courses in natural resources and environmental sciences. He has published frequently in the *Journal of Environmental Education, NAPEC Quaterly*, and has presented scholarly papers at conferences in the United States and the United Kingdom.

Isaiah Smithson is Professor of English Language and Literature at Southern Illinois University at Edwardsville. He was co-editor of *English Studies/ Culture Studies. Institutional Dissent* (1994) and *Gender in the Classroom: Power and Pedagogy*. He is currently at work on a book manuscript titled *American Conceptions of the Forest: Thoreau, Paul Bunyan, and the U.S. Forest Service*.

Martin Stanisstreet is Senior Lecturer at the University of Liverpool, where he teaches biological sciences and is a member of the Environmental Education Research Unit. His research interest is in children's ideas about the environment. He has presented papers on the curricular implications of his findings at scholarly conferences in the U.S. and U.K.

Patricia J. Thompson is Professor in the Division of Education at Lehman College, where she teaches courses in Women's Studies, Human Ecology, and Counseling. She was the founder and first Chair of the Special Interest Group (SIG) Ecological/Environmental Education of the American Educational Research Association and was the convenor of the 1995 conference on which this volume is based. She is the co-author of texts in housing and nutrition and has lectured and published her work on Hestian feminism in the United States, Canada, Germany, Greece, and Finland.

Rosanne Wille is Provost and Senior Vice President for Academic Affairs at Lehman College. She has published extensively in the areas of health care and the impact of pet companionship on health. In addition, she has more than fifteen years of experience in higher education administration.

Nancy E. Wright is a Ph.D. candidate in Political Science at the CUNY Graduate Center. Her special interest is in comparative environmental policy, especially in the United States, Puerto Rico, Brazil, and Japan. She has been editor of *Political Science News*, published by the Graduate Center. She has taught environmental policy courses at Lehman College and at the New School.